Major Appliances

Ron Hazelton, chief consultant for **HOW TO FIX IT**, is the Home Improvement Editor for ABC-TV's *Good Morning America* and host of his own home improvement series, Ron Hazelton's *HouseCalls*. He has produced and hosted more than 200 episodes of *The House Doctor,* a home-improvement series airing on the *Home and Garden Television Network* (HGTV).

On television, and in real life, Ron is a coach who visits people in their own homes, helping them do things for themselves. He pioneered the concept of on-location, home-improvement television, making over 600 televised house calls, doing real-life projects with real people.

The son of a building contractor, Ron has always had a fascination with the home and how it works. He left a successful career as a marketing executive to learn woodworking, eventually becoming a Master Craftsman and cabinetmaker. In 1978, he founded Cow Hollow Woodworks in San Francisco, an antique restoration workshop that restored over 17,000 pieces of furniture during his tenure.

Christopher Lombardi is the founder and owner of *Absolute Repair,* an appliance repair business since 1994. Mr. Lombardi has twenty years of diagnostic, electrical, and mechanical experience, seven of which has been dedicated to troubleshooting and repairing major appliances. He and his staff of specialized appliance technicians sell, install, and repair all major brands of appliances.

Evan Powell is the Director of Engineering and Services at Southeastern Products, Inc. The author of *The Complete Guide to Home Appliance Repair*-and many other widely acclaimed books on home repair, Mr. Powell is also a regular contributor to numerous industry publications. He appears frequently on television as an expert on home repair and consumer topics.

Major Appliances

By The Editors of Time-Life Books, Alexandria, Virginia

With **TRADE SECRETS** From **Ron Hazelton**

Contents

FIX IT: Refrigerators

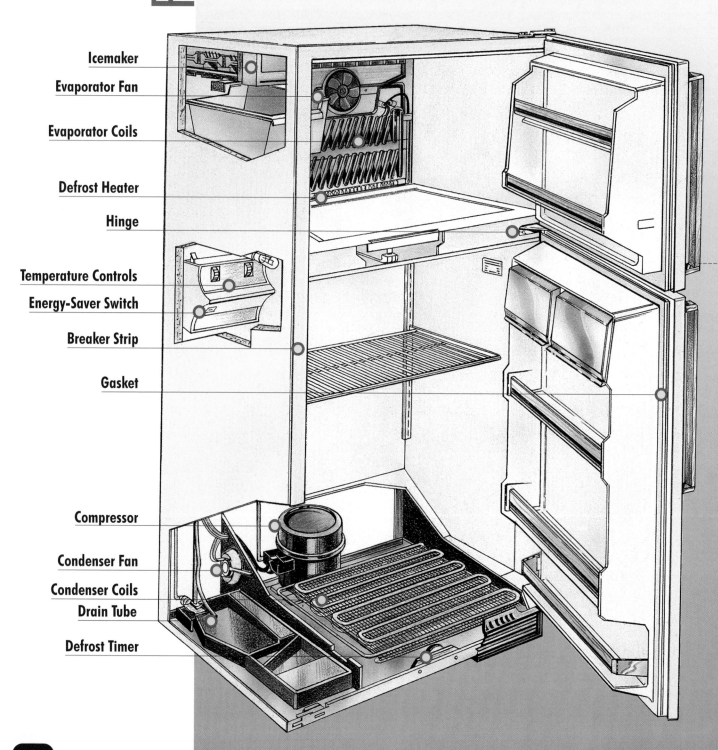

Icemaker

Evaporator Fan

Evaporator Coils

Defrost Heater

Hinge

Temperature Controls

Energy-Saver Switch

Breaker Strip

Gasket

Compressor

Condenser Fan

Condenser Coils

Drain Tube

Defrost Timer

Chapter 1

Contents

How They Work

The illustration at left shows a two-door, frost-free refrigerator with floor-level condenser coils. It contains most features found in typical refrigerators.

Refrigerators work by means of a sealed cooling system. A compressor pumps liquid refrigerant under high pressure through a narrow capillary tube into the evaporator coils. There the liquid quickly boils and expands into a gas, absorbing heat from inside the refrigerator to cool it. Next pressure from the compressor forces the gas to the condenser coils, which dissipate the heat to the outside air. Now a liquid, the refrigerant passes again through the narrow capillary tube as the cycle of heating and cooling continues.

Troubleshooting

Problem	Solution
• **Refrigerator doesn't run, and the light doesn't work**	Check that the refrigerator is plugged in • Examine the power cord for damage • Check for a blown fuse or tripped circuit breaker •
• **Refrigerator doesn't run, but the light works**	Clean the condenser coils **11** • Check the temperature **19** • Test the temperature control **19** • Test the evaporator fan **22** • Service the compressor **29** • Test the defrost timer **25** • Test the compressor relay **29** • Test the overload protector **31** • Service the condenser fan and motor **27** •
• **Refrigerator starts and stops rapidly**	Clean the condenser coils **11** • Service the condenser fan and motor **27** • Test the overload protector **31** • Service the compressor **29** • Have an electrician check the voltage at outlet if overload protector trips frequently •
• **Refrigerator runs constantly**	Defrost refrigerator • Clean the condenser coils **11** • Check the door seal **12** • Replace the door gasket **16** • Service the condenser fan and motor **27** •
• **Refrigerator not cold enough**	Check the temperature **19** • Test the temperature control **19** • Clean the condenser coils **11** • Check the door seal **12** • Replace the door gasket **16** • Test the door switch (some models) **17** • Test the evaporator fan **22** • Test the defrost heater **24** • Test the defrost timer **25** • Call for service if you suspect refrigerant leak •
• **Refrigerator too cold**	Move temperature to higher setting • Check the temperature **19** • Test the temperature control **19** •
• **Refrigerator doesn't defrost automatically**	Service the defrost heater **24** • Service the defrost timer **25** •
• **Ice in drain pan or water in bottom of refrigerator**	Clean the drain tube **12** •
• **Water on floor around refrigerator**	Clean the drain tube **12** • Check the position of the drain pan **12** •

Troubleshooting

Problem	Solution
• **Interior light doesn't work**	Replace the bulb and test the switch **17** •
• **Refrigerator smells bad**	Remove spoiled food; wash interior with baking soda and warm water • Wash the drain pan **12** • Remove the breaker strips and allow the insulation to dry **21** •
• **Moisture on the door and frame**	Inspect the breaker strips; replace if necessary **21** •
• **Refrigerator runs noisily**	Refrigerator not level; adjust leveling feet • Reposition the drain pan **12** • Check evaporator fan **22** • Inspect condenser fan **27** • Replace compressor mountings **29** •
• **Icemaker doesn't make ice**	Check freezer temperature **19** • Test the motor and heater (on modular type) **32** • Install a new module **33** • Check the thermostat **33** • Test the ON/OFF switch (on icemakers with separate controls) **36** • Test the holding switch **36** • Test the water inlet valve switch **37** • Test the motor **37** • Test the thermostat **38** • Clean the water inlet valve filter **39** • Test the water inlet valve **39** •
• **Icemaker won't stop making ice**	Prop up shutoff arm; if ice stops, check arm position • Test the ON/OFF switch (on icemakers with separate controls) **36** •
• **Water on floor**	Check the water inlet valve connections **39** •
• **Water overflows from icemaker**	Level the refrigerator feet • Test the water inlet valve switch **37** • Test the inlet valve **39** •
• **Icemaker doesn't eject ice cubes**	Test the holding switch (non-modular icemaker) **36** • Test the motor **37** • Test the thermostat **38** •
• **Ice cubes discolored or flecked**	Clean the water inlet valve filter **39** • Ice mold worn; call for service • Hard water problem: Install in-line filter; call for service •
• **Ice cubes smell or taste bad**	Clean refrigerator and freezer; flush drain **12** • Remove spoiled food; rewrap food • Wash ice bin • Install in-line filter; call for service •

Before You Start

TOOLS

Cooking thermometer

Basting syringe

Wrench and socket set

Hex wrench

Screwdrivers

Putty knife

Wire brush

Long-nose pliers

Multitester

Pry bar

MATERIALS

Baking soda

Masking tape

Metallic putty

Rags

SAFETY FIRST

Before starting any repair, always unplug the refrigerator or disconnect the fuse or circuit breaker.

Refrigerators are extremely heavy. Never attempt to tilt the refrigerator forward or backward without a helper; the machine can fall on you.

An average life span of 15 years puts refrigerators among the longest lasting and most trouble-free of all major appliances.

KEEP THE MACHINE CLEAN

While a spotless refrigerator isn't a necessity, having a clean one is important not only for appearances, but also for trouble-free operation. Wipe up spills immediately to avoid clogged drain tubes, which can lead to odors and ice build-up. Flush a solution of warm water and baking soda through the drain tubes with a basting syringe. Wash the compartments, trays, shelves, and drain pan twice a year using a water and baking soda solution. To avoid over-heating, regularly wipe dust and dirt from the grille and condenser coils at the bottom or back of the machine.

Refrigerators and freezers are most efficient when they are about three-quarters full. Fill plastic soda bottles or milk jugs with water and place them in the refrigerator to compensate for an excess of empty space.

SOLVING PROBLEMS

You can diagnose and repair poorly sealed doors and gaskets, the door switch, evaporator fan, defrosting system, refrigerator and freezer temperature controls, and the icemaker—all with common tools. However, repairs to the cooling system require special skills and equipment and must be handled by licensed professionals.

Before You Start Tips:

····⫶ Never hammer at ice build-up or attempt to pry it away with a sharp tool; you'll risk puncturing the evaporator coils. Instead, melt ice with a hair dryer, or place pans of hot water in each compartment. Have rags on hand to soak up overflow from the drain pan.

····⫶ If there's a power outage, food will keep in a closed refrigerator for 24 to 36 hours. Keep the door closed as much as possible. If food must be removed from the refrigerator, place it in a bathtub layered with newspaper and ice.

····⫶ If the refrigerator isn't running and the light is out, first check the power cord. It can be damaged or severed if the refrigerator's weight is shifted onto it. Also check for a tripped circuit breaker or blown fuse.

Condenser Coils

1. FLOOR-LEVEL COILS

An accumulation of dirt and dust prevents condenser coils from dissipating heat, making the refrigerator cool poorly, run constantly, or even stop completely if the compressor overheats. Clean floor-level coils twice a year; more often if you have pets.

• Unplug the refrigerator before cleaning the condenser coils.

• Pull off the grille, then use a vacuum cleaner with a wand attachment to remove any dust and pet hair that has accumulated behind it *(left)*.

2. REAR-MOUNTED COILS

Clean rear-mounted coils yearly.

• Use a vacuum cleaner or a stiff brush *(left)*.

• Wash greasy coils with warm, soapy water, taking care not to drip water on other parts of the refrigerator.

Drain and Drain Pan

1. A FLOOR DRAIN

Ice in the drain pan or water pooling inside the refrigerator may indicate a clogged drain. The drain pan is located under the refrigerator behind the front grille. Wash the pan regularly with a warm water and baking soda solution. If the pan rattles, it may be too close to the compressor; reposition it.

Basting Syringe

• Look under a storage drawer to locate the floor drain.

• With a basting syringe, force a solution of warm water and baking soda or bleach into the opening *(right)*.

2. A WALL DRAIN

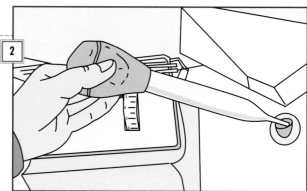

• Pull off the drain trough to expose the drain. With a basting syringe, force a solution of warm water and baking soda or bleach into the opening *(right)*.

• To clear a stubborn clog, insert a length of 1/4-inch round plastic tubing into the drain. Push it through to the drain pan below, then pull it out.

Door Seal

1. EXAMINING THE GASKET

A refrigerator must be level for a tight door seal. Adjust the feet to the refrigerator if necessary.

• Feel the gasket for brittleness and inspect it thoroughly for cracks. Replace a defective gasket *(page 16)*.

• Open the door, place a dollar bill against the breaker strip, then shut the door.

• Slowly pull the bill out *(right)*. If the gasket seals properly, you will feel tension as you pull. Repeat test all around the door.

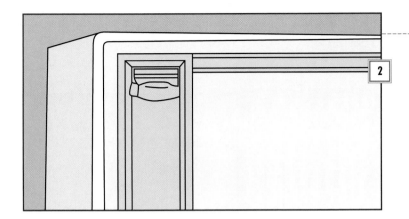

2. EVALUATING DOOR ALIGNMENT

A door that sags on its hinges will result in a poor seal.

- Check the seal between the door gasket and the cabinet for obvious gaps.

- Examine the alignment of the door with the top and sides of the refrigerator cabinet. The top and sides of the door should be parallel to the cabinet. If they are not *(left)*, try adjusting the hinges *(page 14)*.

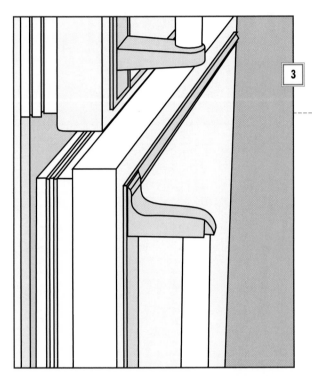

3. CHECKING THE DOOR FOR WARPING

- Check to see if the door gasket lies flat against the refrigerator cabinet *(left)*. Make sure the inner panel isn't rubbing against the breaker strip (often caused by overloading door shelves).

- Check the alignment of the inner and outer door panels. Inspect the outer panel for bowing.

- Try adjusting the hinges *(page 14)*; if this doesn't solve the problem, reposition the door panels *(page 15)*.

REDUCING FROST AND MOISTURE BUILD-UP

Newer refrigerators cool to lower temperatures and have better insulation. Moisture entering the refrigerator or freezer—especially on hot and humid days—condenses to form water or frost inside the appliance. Even frost-free models suffer. Here are a few tips to reduce frost and moisture build-up:

- Open refrigerator and freezer doors as little as possible.
- Arrange shelving and food so air can circulate.
- Tightly seal liquids stored in the fresh food section.
- Use an air conditioner to reduce humidity in your house.
- Avoid over- or underloading the freezer. It should be at least half empty to allow air to circulate between items.
- Use the manufacturer's recommendations for seasonal settings.
- To reduce moisture in the kitchen, use the range hood fan and cover pots and pans during and after cooking.
- Clean the condenser coils.
- Maintain the door gaskets.

Sagging Doors

1. THE UPPER HINGE

Any of three hinges may be adjusted to fix a sagging upper or lower door.

• To adjust a top-mounted freezer door, remove the hinge cover, if present.

• Loosen the bolt on the upper door hinge with a wrench *(right)*, then realign the freezer door with the refrigerator cabinet.

• Hold the door in place and retighten the bolt; replace the hinge cover.

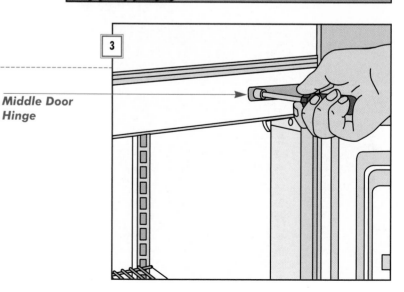

Upper Door Hinge

2. THE LOWER HINGE

• For a slight adjustment to a refrigerator door, first loosen the bolt on the lower hinge with a wrench *(right)*.

• Lift or push the refrigerator door square with the refrigerator cabinet, then retighten the bolt.

• If adjustment doesn't hold, check to see if the hinge cam is worn. Replace it if needed.

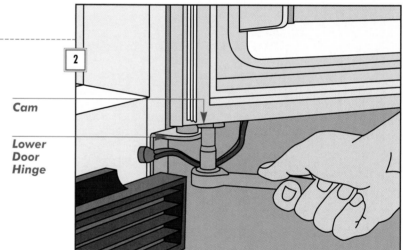

Cam

Lower Door Hinge

3. THE MIDDLE HINGE

• Locate the middle hinge between the door and face of the refrigerator cabinet.

• Loosen the bolts with a wrench or nut driver *(right)*.

• Shift the hinge in or out to realign the door; retighten the hinge bolts.

• If the door still doesn't seal properly, realign the door panels to correct warping *(opposite)*.

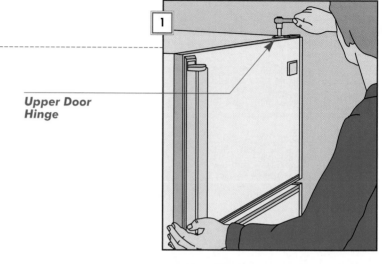

Middle Door Hinge

Warped Doors

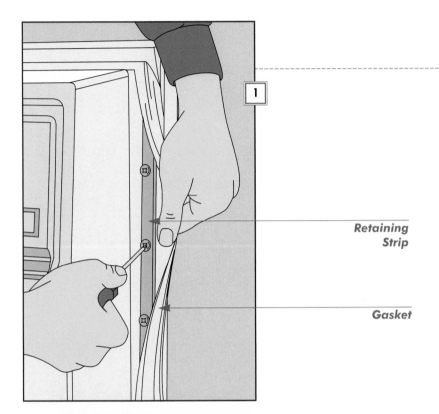

1. LOOSENING THE RETAINING SCREWS

A misaligned door panel will result in a poor seal.

• Open the refrigerator door and pull back the gasket to expose the interior door panel's retaining strips.

• Loosen, but do not remove, all of the screws along all four sides of the panel *(left)*.

Retaining Strip

Gasket

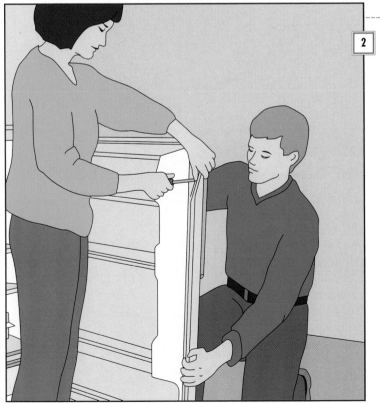

2. ALIGNING THE DOOR PANELS

• Grasp the outer door panel at the top and side, and twist it opposite to the warp.

• Hold the door in this position; have a helper partially retighten the screws *(left)*.

• Close the door and check to see if the warp has been corrected.

• If the door seals properly, reopen it and hold it firmly in position while a helper retightens the screws.

• Avoid placing heavy items on the door shelves. Replace the door if the warping persists.

Door Gaskets

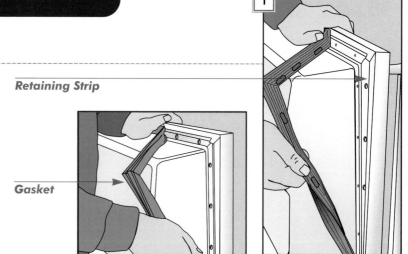

1. REMOVING THE OLD GASKET

• Pull back the gasket to expose the retaining strip and loosen the screws *(page 15)*.

• If it is merely clamped to the door by the retaining strip, pull the gasket free *(near right)*.

• If the screws pass through the gasket, remove the screws and gasket along the top and one-third of the way down both sides of the door *(far right)*.

Retaining Strip

Gasket

2. ATTACHING A NEW GASKET

• With a helper holding the door to prevent warping and starting at an upper corner, insert the gasket flange behind the retaining strip. If screws pass through the gasket *(Step 1)*, fasten the gasket to the top of the door and along the top-third of the sides *(right)*. Then, attach the middle third followed by the lowest third and the bottom of the door.

• Partially tighten screws as you progress.

• After the gasket is in place, tighten screws securely, starting at the bottom corners.

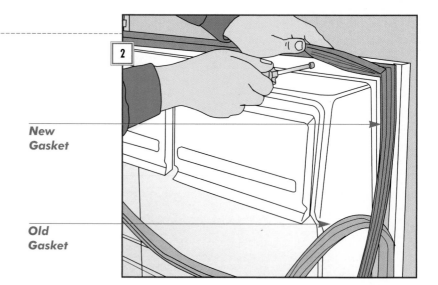

New Gasket

Old Gasket

Ron's TRADE SECRETS

CLEANING GASKETS IN PLACE

Keep refrigerator gaskets clean so they don't become brittle and crack. Whenever I notice mildew starting to form, I use an old toothbrush and some bleach to scrub between the gaskets' accordion-like folds. Be sure not to pull too hard when you're cleaning a gasket—you can easily rip or crack it, turning a simple repair into a more difficult one. To prevent mildew, check your refrigerator's door seals *(pages 12-13),* and turn off the energy-saver switch to reduce moisture build-up.

Door Switch

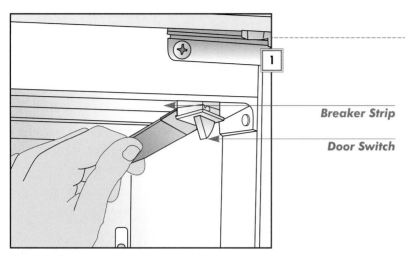

Breaker Strip

Door Switch

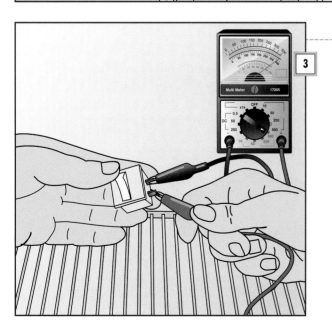

1. REMOVING THE DOOR SWITCH

In most refrigerators, a single switch controls the interior light. If the light doesn't work, first try replacing the bulb. If the new bulb doesn't glow, remove and test the switch. If you suspect that the light stays on when the door is closed, press the switch by hand. If the light doesn't go out, remove and test the switch.

● Unplug the refrigerator.

● Pry out the switch with a putty knife; pad the blade with masking tape to avoid scratching the breaker strip *(left)*.

2. DISCONNECTING THE SWITCH

● Ease the switch out of its recess, then pull on it to expose a few inches of wiring.

● The switch will have either two or four terminals; four terminals indicate that it is a combination fan and light switch.

● Remove the push-on connectors *(left)*, and label the wires for reassembly.

● Splice on new connectors if the existing ones are burned or corroded *(page 51)*.

3. TESTING THE SWITCH

● Set a multitester to RX1 *(page 18)*.

● If the switch has two terminals, clip the probe to one terminal and touch the second probe to the other terminal *(left)*. If the meter indicates continuity with the switch button out, and an open circuit with the button in, the switch is good.

● If the switch fails any test, replace it.

HOW TO USE A MULTITESTER

A multitester is an inexpensive but invaluable device for diagnosing problems in electrical equipment. Although the instrument can measure voltage, the repairs in this book only use it to measure resistance. As described below, one test is for continuity, another for determining the resistance of a component (expressed in a unit called *ohms*), and a third to tell whether a circuit or electrical component is grounded. All three tests are safely done with the power off.

Multitesters come in two styles: The analog type, shown here, has a needle indicator and printed scales that correspond to the settings of a rotary switch used to choose the test mode. Digital multitesters give numerical readouts according to the test mode selected.

All multitesters have probes connected to positive and negative leads. Alligator clips can be slipped over the probes and clipped to terminals when convenient.

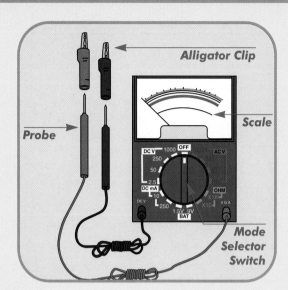

Alligator Clip

Scale

Probe

Mode Selector Switch

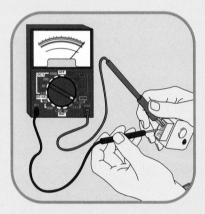

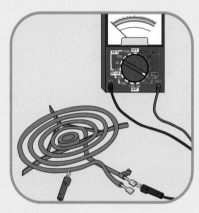

TESTING FOR CONTINUITY

A continuity test determines if a circuit is complete or open. Always unplug the appliance and disconnect the component being tested. Set the dial to RX1 and touch probes to terminals.

The illustration above shows a continuity test on a switch. With the switch ON and probes touching two terminals, the needle indicates 0 resistance—a complete circuit. An infinity reading indicates an open circuit.

Routinely "zero" analog meters by touching the probes together and adjusting the ZERO OHMS dial until the needle falls directly over 0 on the scale.

MEASURING RESISTANCE

A resistance test determines how much the flow of electricity in a circuit is impeded—an important measurement for specialized circuits like those in electric range heating elements, which depend on resistance to heat up. Unplug the appliance and disconnect all leads to the component being tested. Zero the meter, then turn the dial to the RX setting specified for the component.

To test a heating element for resistance *(above)*, touch a probe to each of the terminals. A low reading (below 120 ohms) is normal. An infinity reading would indicate an open circuit.

TESTING FOR GROUND

A test for a ground fault determines whether there is any potentially dangerous leakage of current from a circuit to another component of an appliance. Ground faults may not only cause an appliance to malfunction, but may also present a shock hazard to anyone using it.

To test a range heating element for a ground fault *(above)*, select the RX1 scale in the ohms function. Touch one probe to one of the terminals and the other to the element's sheathing. On an analog-type tester, if the needle moves at all, it indicates a potentially dangerous ground fault.

Temperature Controls

1. CHECKING TEMPERATURES

The ideal refrigerator temperature is between 38°F and 40°F; for the freezer compartment, between 0°F and 8°F (the freezer temperature may be about 10°F higher on single-door models).

• Place a cup of water in the refrigerator and leave for 2 hours *(left);* use cooking oil in the freezer.

• Place a cooking thermometer in the liquid and wait 3 minutes.

• Adjust the temperature control accordingly; if the problem persists, test the temperature control.

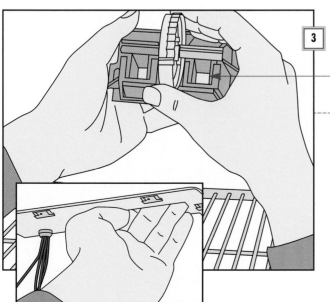

Freezer Vent Control

Temperature Control

Energy-Saver Switch

Console

2. REMOVING THE CONTROL CONSOLE

• Unplug the refrigerator.

• Unscrew and remove the console that houses the temperature control and the energy-saver switch *(left).*

• Let the console dangle by its wiring; the freezer vent control will remain attached to the refrigerator wall.

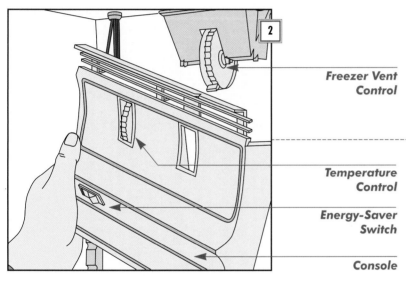

Louvered Vent

3. CHECKING THE FREEZER CONTROL

• Unscrew the freezer vent control from the refrigerator wall.

• Use a hair dryer on LOW to melt any ice.

• Remove any food residue in the vent.

• Reach into the freezer channel and check for obstructions *(inset).*

• Reinstall the freezer vent control.

4. Testing the energy-saver switch and temperature control

• Remove the push-on connector and the ground wires from the temperature control terminals *(right)*.

• Set a multitester to test for continuity *(page 18)*.

• With the temperature control turned to its coldest setting, the tester should read 0 ohms; with the control turned off, it should read infinity.

• Replace a faulty control *(next step)*.

• Remove the push-on connector on the energy-saver switch (if present) and test the switch for continuity *(inset)*.

• Place a probe on each terminal; the tester should show continuity when the switch is on.

• If the switch is faulty, pull it off the console, snap on a new switch, and reattach the push-on connector.

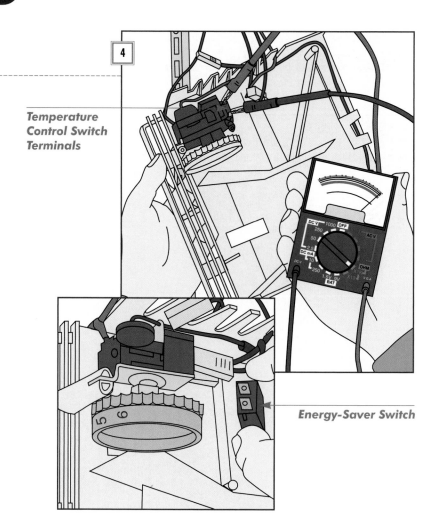

Temperature Control Switch Terminals

Energy-Saver Switch

5. Replacing the temperature control

• Note the position of the temperature control's capillary line in the console; the new one must be installed the same way.

• Pull out the old control and snap in a new one, threading the capillary line into place in the console *(right)*.

• Remount the console on the refrigerator wall; recheck the temperature *(page 19)*.

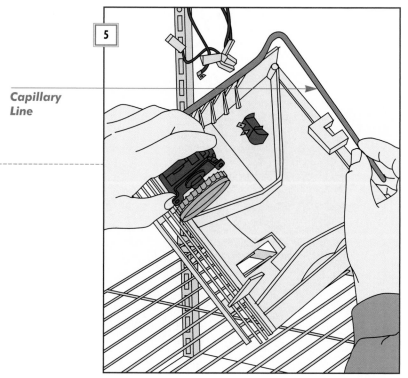

Capillary Line

Breaker Strips

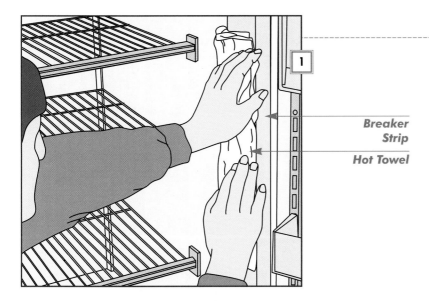

Breaker
Strip

Hot Towel

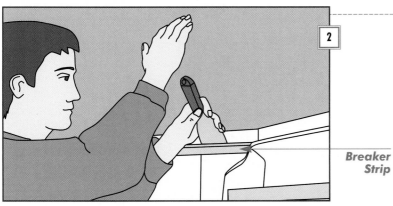

Breaker
Strip

1. SOFTENING A BREAKER STRIP

Damaged breaker strips allow moisture to enter the insulation between the inner and outer walls of the refrigerator.

• Inspect the breaker strips around the inner frame for warps and cracks.

• To replace a damaged breaker strip, first unplug the refrigerator.

• The breaker strip is brittle when cold. To prevent snapping it during removal, first soften the strip by pressing a hot, wet towel along its entire length *(left)*.

2. RELEASING THE STRIP

• With a putty knife, pry the breaker strip away from the cabinet *(left)*. Be careful not to damage the foam insulation behind the breaker strip.

• If the insulation is damp or ice-clogged, or if it has an unpleasant odor, leave it uncovered for a few hours or dry it with a hair dryer set on LOW.

• Snap a new breaker strip in place.

REMOVING OTHER TYPES OF BREAKER STRIPS

Breaker strips are installed differently from model to model. Some are attached with sealant: Cut through sealant with a utility knife. Replace the strip and reseal the corners with an arsenic-free sealant rated for use in food compartments.

The most difficult breakers to remove run the full height of the appliance, from the bottom of the refrigerator to the top of the freezer, and are held in place by the center console: Remove the screws that hold the console in place. Pull the console forward and rest it in the freezer without disconnecting the wires *(right)*. Replace the breaker strip *(above)* and reinstall the console.

Evaporator Fan

1. EXPOSING THE EVAPORATOR COVER

The evaporator coils are part of the sealed refrigeration system and should only be serviced professionally. But the evaporator fan and the defrost heater—usually located in the rear wall of a top-mounted freezer behind the evaporator cover—are easy to test and replace.

• Unplug the refrigerator.

• Remove any obstructions that prevent access to the rear wall of the freezer, including the freezer shelves *(right)* and icemaker, if present *(page 34).*

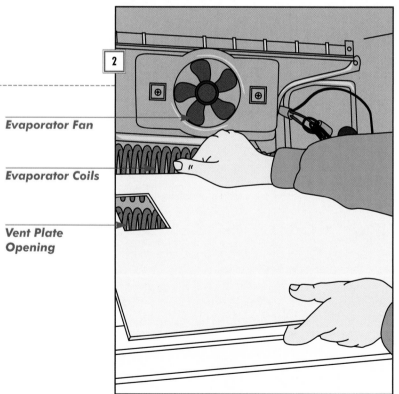

Icemaker Mount

2. REMOVING THE EVAPORATOR COVER

• Remove the screws from the vent plate, as well as any brackets that may secure the back wall of the freezer.

• Pull off the back wall of the freezer compartment to expose the evaporator coils, evaporator fan, and the defrost heater *(right).*

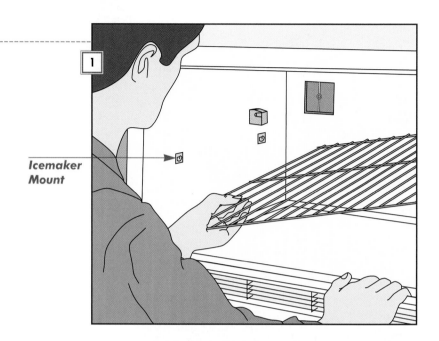

Evaporator Fan

Evaporator Coils

Vent Plate Opening

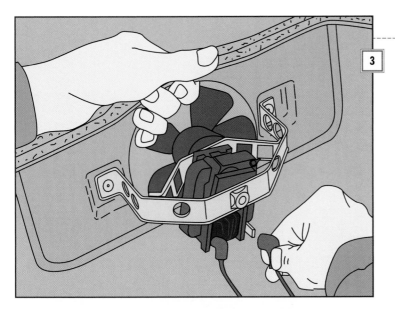

3. FREEING THE FAN ASSEMBLY

• If present, remove the screws that attach the evaporator fan housing.

• Lift the fan assembly a few inches and disconnect the wires *(left)*. Use long-nose pliers if you cannot pull off the wire connectors with your fingers.

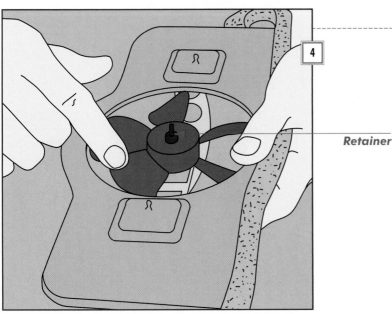

Retainer

4. CHECKING THE FAN BLADE

• If the blade shows signs of damage, pull off the retainer at the center, pull the blade off the motor shaft, and slide on the new blade. Be sure to orient the angle of the blade fins in the same direction as the old blade and replace the retainer.

• Hold the fan horizontally in one hand, then spin the blade to check for binding in the motor *(left)*.

• Replace the motor if the blade does not spin freely *(next step)*.

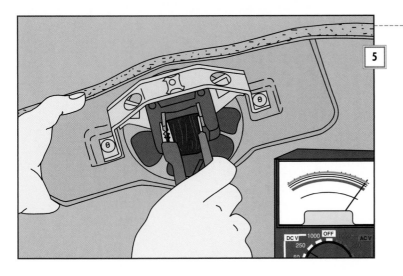

5. TESTING THE MOTOR

• Set a multitester to RX1. Touch a probe to each motor terminal *(left)*. The meter should show continuity *(page 18)*; if not, replace the motor.

• To remove the motor, first remove the fan blade. Unscrew the smaller bracket at the front of the motor, and remove the motor from the housing.

• Install the new motor and replace the bracket and fan blade (take care not to reverse the blade).

Defrost Heater

1. TESTING A DEFROST HEATER

The defrost heater element may be enclosed in a glass tube hidden beneath a metal shield mounted under the evaporator coils *(top right)*. Alternatively, it may be wrapped in aluminum foil *(top inset)*, or it may be an exposed metal rod *(bottom inset)*. Test all elements the same way.

• Unplug the refrigerator.

• Remove the evaporator cover *(page 22)*.

• Pull the wire connectors from the terminals at each end of the defrost heater.

• Set a multitester to RX1 and attach a probe to each defrost heater terminal to test for continuity *(page 18)*.

• If the test indicates that the circuit is open—infinite resistance—replace the defrost heater *(next step)*.

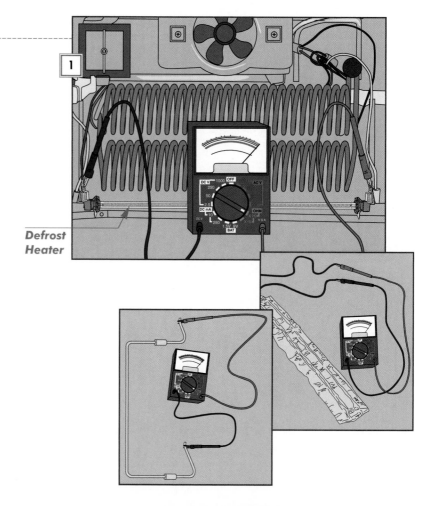

Defrost Heater

2. REMOVING THE DEFROST HEATER

• Locate and unhook the element's reflector shield from the clips located at each end *(right)*.

• Carefully lift the element out of its brackets.

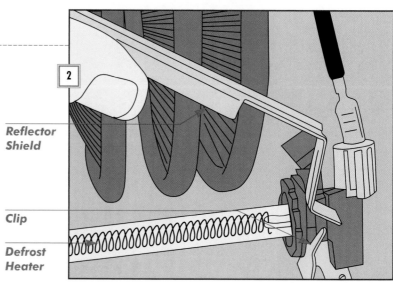

Reflector Shield

Clip

Defrost Heater

3. CLEANING THE DRAIN TUBE

● Clear the drain tube while you have access to it: With a basting syringe, force a solution of hot water and baking soda or bleach into the opening.

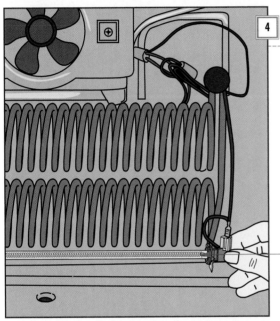

Defrost Heater Element

4. INSTALLING THE NEW DEFROST HEATER

● Do not touch the glass surface of the new heater element; oils from your skin will cause hot spots.

● If you do touch it, wipe the element thoroughly with a paper towel moistened with rubbing alcohol.

● Plug the new heater element in the same position as the old one and replace all clips or fasteners *(left)*.

● Reconnect the push-on connector.

● Reinstall the evaporator cover.

Defrost Timer

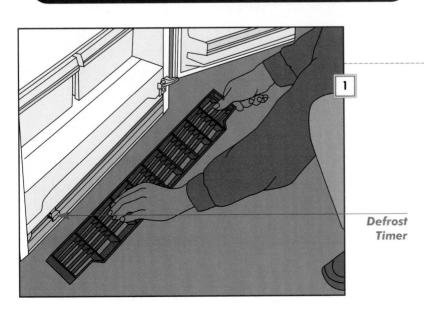

Defrost Timer

1. REMOVING THE DEFROST TIMER

The defrost timer is usually located behind the front grille of the refrigerator. Alternatively, it may be located in the compressor compartment at the back of the refrigerator, in the thermostat console, or behind a cover plate inside the refrigerator.

● Unplug the refrigerator.

● Remove the front grille *(left)*.

● Unscrew the timer from its mounting plate and slide it out.

2. DISCONNECTING THE WIRES

- After unscrewing the defrost timer, gently wiggle it out of the plug connecting it to the refrigerator *(right)*. The plug is polarized so that you cannot reconnect the defrost timer incorrectly.

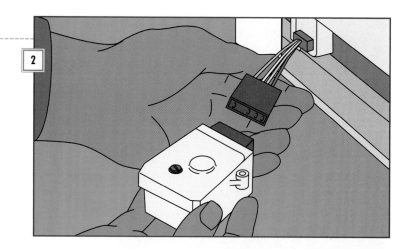

3. TESTING THE TIMER

- Find the common terminal of the timer. It is usually connected to the white wire of the harness plug; if the terminals are numbered, it is number 3.

- Set a multitester to RX1.

- Attach one multitester probe to the common terminal and touch the other probe to each of the three other terminals *(top right)*.

- Two of these terminal pairs should have full continuity, while the third should have no continuity *(page 18)*.

- Turn the defrost timer switch clockwise until you hear a click *(bottom right)*.

- Test the timer again. Two of the terminal pairs should show continuity, while one—not the same pair as before—should not.

- Replace the timer if it fails either test.

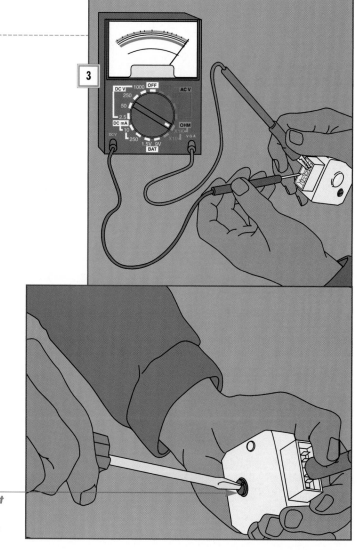

Defrost
Timer
Switch

Condenser Fan

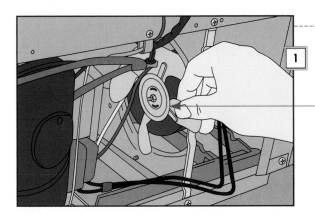

Condenser
Fan

1. INSPECTING THE CONDENSER FAN

• Unplug the refrigerator and pull it away from the wall.

• Remove the access panel, if any.

• Clean the fan blade and turn it to see if the blade rotates freely *(left)*.

• To replace a damaged blade, unscrew the nut that holds it to the motor, then pull it off. Install a new fan blade, replacing any washers, and tighten the nut *(page 28)*.

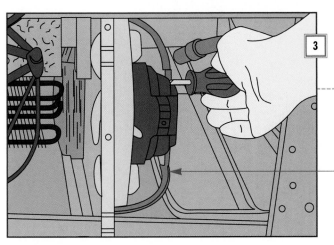

2. TESTING THE FAN MOTOR

• Disconnect the wires to the fan motor.

• Set a multitester to RX10 and touch one probe to each terminal *(far left)*.

• The multitester needle should move to the medium range of the scale, indicating partial resistance; a reading of 0 or infinity means the motor is faulty and should be replaced *(next step)*.

• Set the multitester to RX1000 and touch one probe to the motor terminals and the other to any unpainted metal part of the refrigerator *(near left)*.

• If the multitester needle moves at all, the motor is grounded and should be replaced.

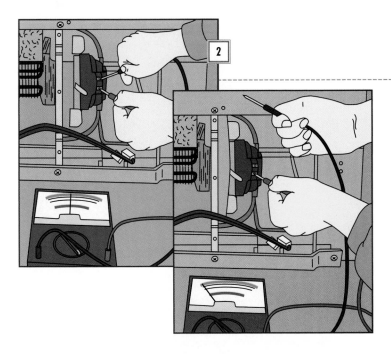

Fan
Bracket

3. REMOVING THE MOTOR

• Unscrew the brackets that hold the fan motor to its housing *(left)*; if necessary, unscrew the mounting plate from the motor.

• Slide the fan motor backwards out of the housing.

4. REPLACING THE MOTOR

• Remove the fan blade from the old motor *(Step 1)* and attach it to the new motor *(right)*.

• Screw the brackets in place to install the new motor in its housing.

• Reattach the wires to the motor terminals.

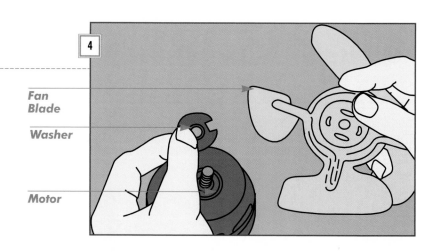

Fan Blade

Washer

Motor

Compressor Mountings

1. LOOSENING THE MOUNTINGS

Excessive compressor vibration causes a refrigerator to run noisily. Compressors are mounted on flexible shock absorbers designed to dampen the vibrations.

• To reach the compressor, pull the refrigerator away from the wall and unplug it.

• Remove the rear access panel, if any.

• With a wrench, unscrew the nut securing one foot of the compressor *(right)*.

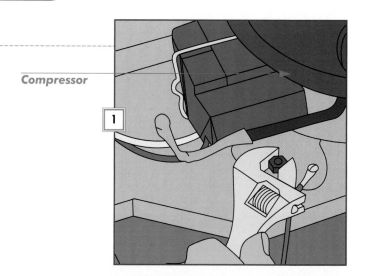

Compressor

2. PULLING OUT THE SHOCK ABSORBERS

• With a pry bar, jack up the compressor foot just far enough to lift out the shock absorber beneath it *(right)*.

• Slip a new shock absorber in place and lower the compressor foot.

• Retighten the nut.

• Repeat for each mounting; remove only one mounting at a time.

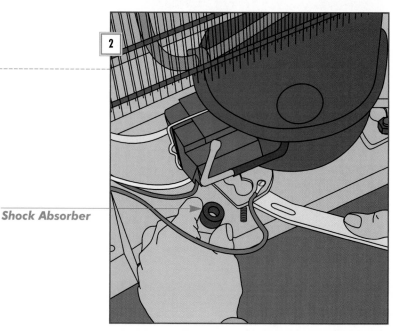

Shock Absorber

Compressor

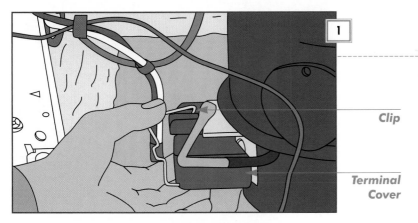

1

Clip

Terminal
Cover

1. REMOVING THE COMPRESSOR TERMINAL COVER

The terminal cover is a small box mounted on the compressor cover, if any, that protects the relay, overload protector, and capacitor.

● Release the wire retaining clip that holds the cover in place. Slip off the cover and the clip *(left)*.

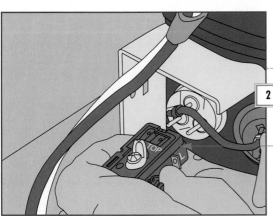

2

Compressor
Relay

2. REMOVING THE COMPRESSOR RELAY

● Pull the relay straight off the compressor without twisting it *(left)*. Remove the wire from the side terminal.

CAUTION: If the compressor has a capacitor, discharge it by placing one lead of a 20-ohm wire-wound resistor on each capacitor terminal before proceeding.

Ron's TRADE SECRETS

THE FACTS ON FREEZERS

Upright and chest-style freezers have a lot in common with refrigerators. Troubleshooting procedures for the thermostat, evaporator fan, and compressor (usually found in freezers behind a removable panel) are pretty much the same.

When a freezer runs constantly, I start by checking the alignment of the freezer door or lid. If it's out of whack, I look at the hinges to see if I can correct the problem.

Chest-style freezers may have one of two types of hinges: Piano hinges—two long knuckles joined by a pivot rod—are impossible to adjust. But correcting lid alignment in a freezer with hinges like the one shown here is easy. Screws joining the hinge to the freezer mount through oblong holes, allowing left-to-right and up-and-down adjustments. I usually loosen all the screws a bit, shift the lid into side-to-side alignment, and let gravity take care of the rest before retightening.

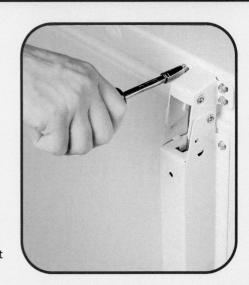

3. TESTING THE COMPRESSOR RELAY

You may find one of two types of compressor relays. If the relay has an exposed wire coil, test it for continuity following the steps below. If there's no visible coil, the relay is of solid-state design and can't be tested for continuity; go on to other compressor tests or call for service.

- Set a multitester to RX1. With the relay upside down *(right),* slip the probes into the terminals marked S and M; the needle should sweep across the scale, showing continuity *(page 18).*

- Remove the probe from M and place it on the side terminal marked L; the needle should again move to 0.

- Remove the probe from S and place it in M; the needle should not move.

- Turn the relay over and listen for a click as the magnetic switch inside the relay engages.

- Perform the same tests. You should get the opposite results: no continuity between terminals S and M, and S and L; continuity between M and L.

- Replace the relay if it fails any of the above tests.

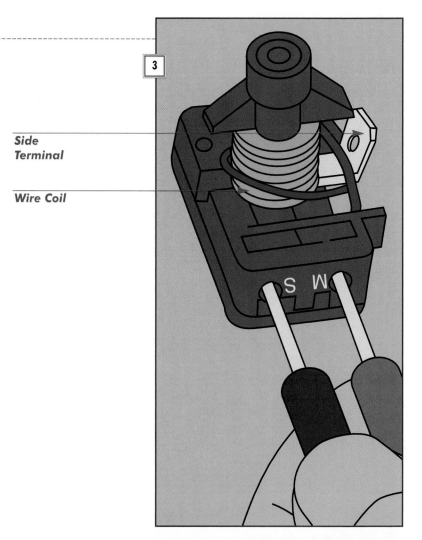

Side
Terminal

Wire Coil

4. REMOVING THE OVERLOAD PROTECTOR

- With a screwdriver, gently pry open the circular spring clip that secures the overload protector to the compressor, then snap out the protector *(right).*

- Pull the two wire connectors off their terminals.

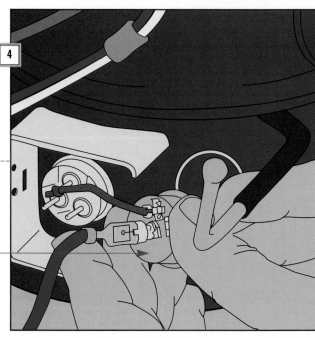

Overload
Protector

5

5. TESTING THE OVERLOAD PROTECTOR

• Set a multitester to RX1, then touch a probe to each overload protector terminal *(left)*.

• The needle should sweep across the scale to 0, showing full continuity.

• Replace the overload protector if it fails this test by reattaching the push-on connector to the new overload protector, clipping it in place, and replacing the terminal cover.

• If the overload protector passes this test, test the compressor *(next step)*.

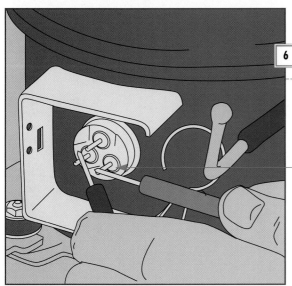

6

Terminal Pin

6. TESTING THE COMPRESSOR MOTOR

• Set a multitester to RX1 and test each of the three terminal pins against the others *(left)*; each pair should show continuity.

• If any pair of terminals fails the test, the compressor motor is faulty; call for service.

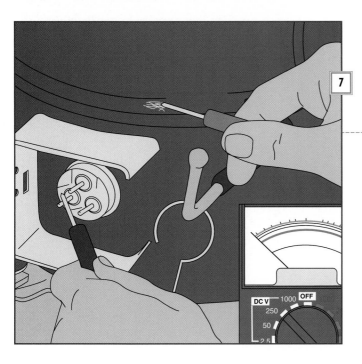

7

7. TESTING THE COMPRESSOR FOR GROUND

• Set a multitester to RX1 and place one probe against the metal housing of the compressor *(left)*. Scrape off a little paint to ensure contact with bare metal.

• Test each compressor terminal. If any of the three terminals shows continuity with the housing, the compressor is grounded; call for service.

Modular Icemaker

How They Work

An icemaker draws water through a water inlet valve, located behind the refrigerator, that taps into the household water supply. The valve meters in enough water to fill the ice cube mold. A thermostat inside the icemaker determines when the water is frozen and activates the motor and mold heater (if any) to eject the ice into a bin. A shutoff arm turns off the icemaker as it is pushed upward by ice accumulating in the bin. The motor and heater in this kind of icemaker are combined in a module that must be replaced as a unit if any part of it malfunctions. However, the thermostat, installed in the back of the icemaker housing, is a separate part. You can replace it if it fails.

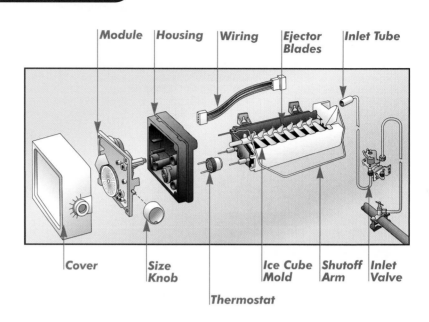

Module Housing Wiring Ejector Blades Inlet Tube

Cover Size Knob Thermostat Ice Cube Mold Shutoff Arm Inlet Valve

Testing the Motor and Heater

A legend and diagram on the back of the icemaker cover show test points for each component on the face of the module *(inset)*. Test the icemaker with the shutoff arm down and the ejector blades in their resting position.

• Unplug the refrigerator, then pry off the icemaker cover with a coin.

• To test the motor, set a multitester to RX1 and insert the probes in test holes L and M *(right)*; the tester should show continuity *(page 18)*.

• Test the heater by placing the probes in holes L and H; the needle should register about 70 ohms.

• Replace the module if either the motor or heater fails the test *(next step)*.

REPLACING THE MODULE

• With the refrigerator unplugged, remove the screws that secure the module, then pull the module off *(left)*.

• Install a new module.

Module

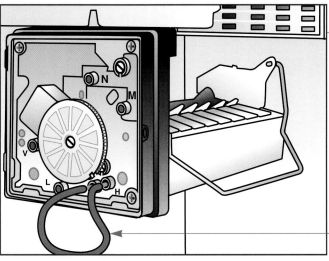

CHECKING THE THERMOSTAT

Test the thermostat with the freezer cold.

• Unplug the refrigerator.

• Insert the bare ends of an insulated wire into holes T and H *(left)*.

• Plug the refrigerator back in (don't touch the wire).

• Replace the thermostat if the motor runs.

Insulated Wire

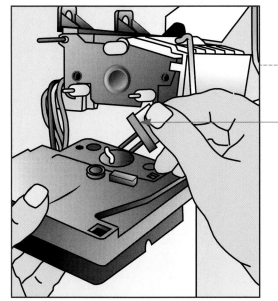

INSTALLING A NEW THERMOSTAT

• Unplug the refrigerator and remove the module *(above)*.

• Remove the screws that secure the housing and detach it from the ice cube mold.

• Slip the old thermostat out of the housing *(left)*.

• Install a new thermostat, using metallic putty where the thermostat touches the housing, as specified by the manufacturer.

Thermostat

Non-Modular Icemaker

HOW THEY WORK

Although it makes ice in the same way as a modular unit *(page 28)*, a non-modular ice-maker *(right)* has a motor and switches—as well as a thermostat—that you can test and replace individually. There are usually three switches: an ON/OFF switch operated by the shutoff arm; a holding switch that keeps power running to the ejector blades in the ice release phase; and a water inlet valve switch that manages the flow of water into the icemaker.

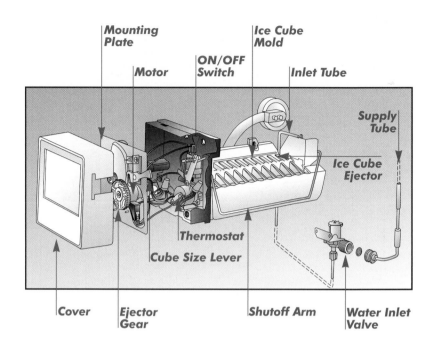

Icemaker Internal Parts

1. DISCONNECTING THE ICEMAKER

● Unplug the refrigerator and set the ice bin aside.

● Remove the screw from the icemaker's bottom bracket.

● Support the icemaker with one hand and remove the top screws and clips *(top right)*.

● Lower the icemaker and unplug it from the refrigerator wall *(bottom right)*.

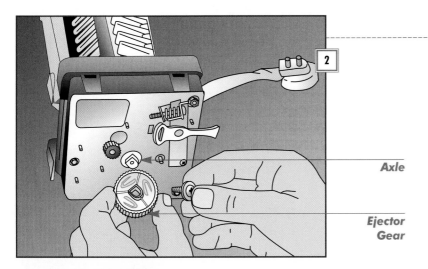

2. INSPECTING THE EJECTOR GEAR

- Use a coin to pry off the cover.

- Inspect the ejector gear for teeth that are worn down or broken.

- Replace the gear if it is damaged. Remove the screw at its center and pull the gear off its axle *(left);* save the washer.

- Place a new gear on the axle and refasten the washer and screw.

Axle

Ejector Gear

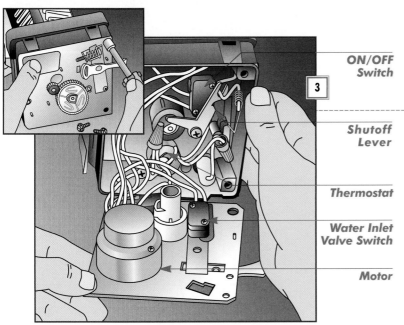

ON/OFF Switch

3. REMOVING THE MOUNTING PLATE

- Take out the screws around the edges of the mounting plate *(inset)* to gain access to the motor, switches, and thermostat, as well as the shutoff lever and arm *(left).*

Shutoff Lever

Thermostat

Water Inlet Valve Switch

Motor

PREVENTING AN ICE CUBE OVERFLOW

If your icemaker runs constantly, causing ice cubes to spill out of the bin, first check to make sure you have the correct bin to catch the cubes. If you are using a different bin or if it is simply out of position, the shutoff arm may not be raised by the build-up of ice cubes, so it cannot turn off the icemaker. The arm extends through the housing that holds the ON/OFF switch and attaches to a short lever that trips the switch when the bin is full. Reposition a shutoff arm that is bent or dislodged from the lever *(right).* If the shutoff arm and tray are in good shape, test the icemaker ON/OFF switch *(page 36).*

ON/OFF and Holding Switches

1. REMOVING THE SWITCHES

Although the ON/OFF and holding switches perform different functions, they are mechanically identical and are tested in the same way. The ON/OFF switch, located behind the shutoff lever, turns off the icemaker when the shutoff arm is raised *(right)*. The holding switch, which is mounted next to the motor, keeps the power going while the ejector pushes out the ice cubes *(inset)*.

• Remove the screws securing the switch.

• Tag the wires with masking tape so that you can reattach them correctly, then pull the connectors off the terminals.

• Unplug the refrigerator, take out the icemaker *(page 34)*, and remove the mounting plate *(page 35)*.

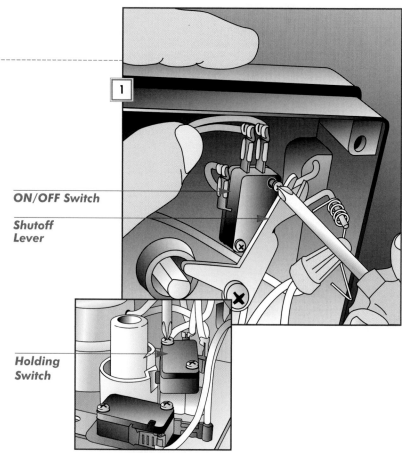

ON/OFF Switch

Shutoff Lever

Holding Switch

2. TESTING THE SWITCHES

• Set a multitester to RX1 and clip one probe to the common terminal on the side of the switch.

• Touch the other probe to each of the other two terminals in turn *(right)*.

• With the switch button out, the tester should show continuity through one terminal and resistance through the other *(page 18)*.

• Press the button in; the results should be reversed.

• If either switch fails either test, replace it.

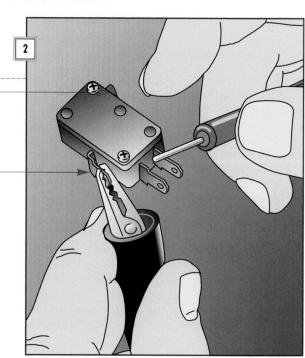

Switch Button

Common Terminal

Water Inlet Valve Switch

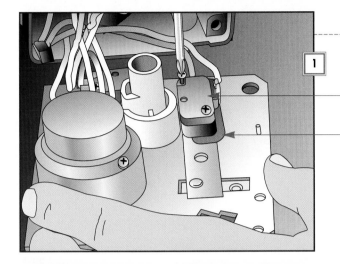

Inlet-Valve
Switch

Insulator

1. REMOVING THE SWITCH

- Unplug the refrigerator, take out the icemaker *(page 34)*, and remove the mounting plate *(page 35)*.

- Unscrew the switch from the plate *(left)*; save the card-like insulator underneath it.

- Disconnect the wires.

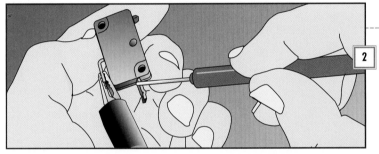

2. TESTING THE SWITCH

- Set a multitester to RX1 and clip one probe to each switch terminal *(left)*.

- With the button out, the tester should show continuity; with the button in, resistance *(page 18)*.

- Replace the switch if it fails the test.

Icemaker Motor

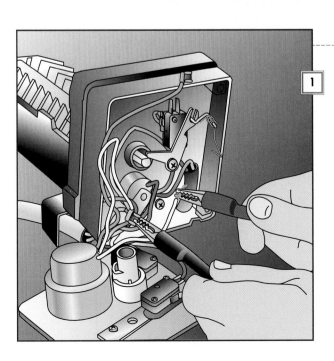

1. TESTING THE MOTOR

- Unplug the refrigerator, take out the icemaker *(page 34)*, and remove the mounting plate *(page 35)*.

- Disconnect two of the three motor wires from the ON/OFF switch, then unscrew the wire cap that joins the third motor wire to a lead from the power supply.

- Set a multitester to RX1 and clip one probe to the third motor wire. Then touch the other probe to each of the two remaining wires in turn *(left)*. A reading of 0 or infinity from either wire means a faulty motor; install a new one *(next step)*.

- If the motor passes the test, check the thermostat *(page 38)*.

2. INSTALLING THE MOTOR

• Unscrew the old motor from its mounting plate *(right)*.

• Install a new motor; make sure the small gear on the motor shaft meshes with the ejector gear on the mounting plate.

• Reconnect the wires, then reassemble and reinstall the icemaker.

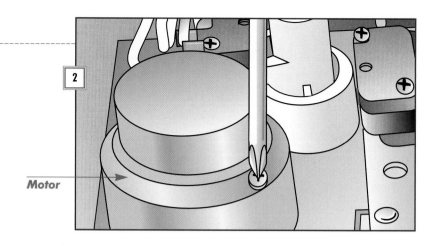

Motor

Icemaker Thermostat

1. REMOVING THE THERMOSTAT

• Unplug the refrigerator, take out the icemaker *(page 34),* and remove the mounting plate *(page 35).*

• Label the thermostat wires, then disconnect them.

• Loosen the clamp that holds the thermostat and remove it *(right).*

Thermostat

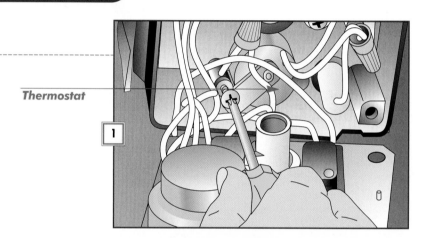

2. TESTING THE THERMOSTAT

• Let the thermostat warm to room temperature. Set a multitester to RX10. If the thermostat has three wires, clip one probe to the shorter of the two wires that end with wire caps. Touch the other probe to the other wires in turn *(right).* A three-wire thermostat should show continuity through one wire and resistance through the other; look for an infinity reading if testing a two-wire thermostat.

• Chill the thermostat in the freezer for 15 minutes and test again. The continuity test results should reverse on a three-wire thermostat; look for a 0 reading on a two-wire thermostat.

• Replace thermostat if it fails either test.

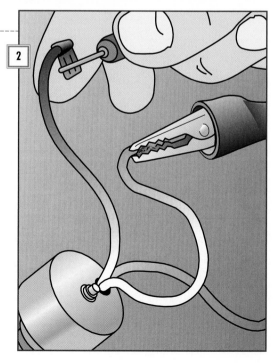

Water Inlet Valve

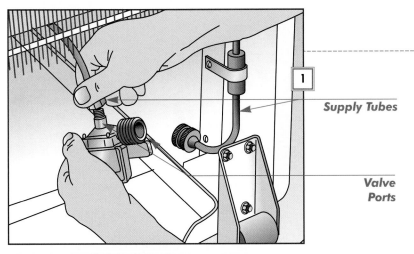

Supply Tubes

Valve Ports

Filter

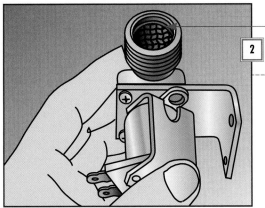

Solenoid Terminals

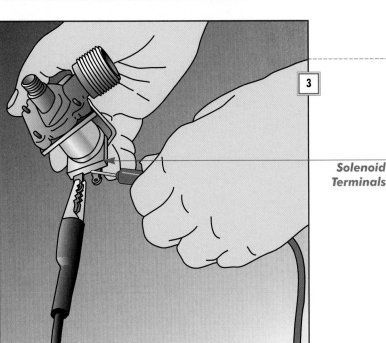

1. REMOVING THE VALVE

• Unplug the refrigerator and turn off the water supply to the icemaker (usually at a valve under the sink).

• Pull the refrigerator out from the wall.

• Unscrew the water inlet valve from its bracket on the back of the refrigerator and disconnect the water-supply tubes from the valve ports *(left)*. Keep a pan handy to catch dripping water.

• Disconnect any wires to the valve.

2. INSPECTING THE FILTER

• Look at the filter screen inside the valve port *(left)*. If it's clogged, try to clean it by placing the whole valve under running water. Don't attempt to remove the filter. If cleaning the screen as directed doesn't work, consider replacing the valve.

3. TESTING THE VALVE

• To test the water inlet valve solenoid, set a multitester to RX100 and touch a probe to each terminal *(left)*.

• The tester should show a reading of 200 to 500 ohms *(page 18)*. Replace the water inlet valve if it fails the test.

• Hook up the terminal plug and ground wire to the new valve, screw the bracket to the refrigerator, and connect the supply tubes to the valve ports.

• Push the refrigerator against the wall, plug it in, and turn on the water.

• After installing a new water inlet valve, discard the first two or three batches of ice.

FIX IT: Electric Ranges

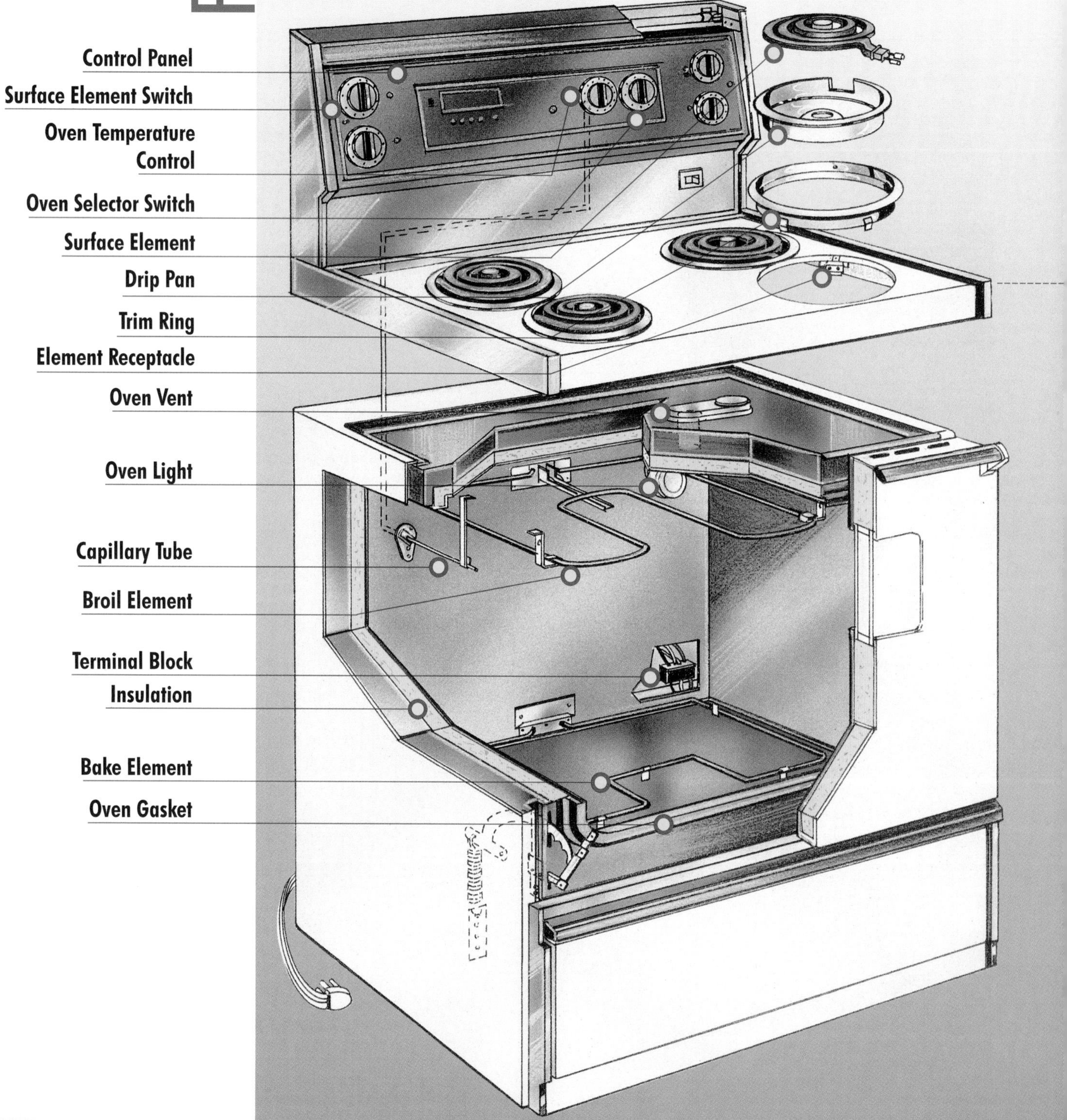

Control Panel

Surface Element Switch

Oven Temperature Control

Oven Selector Switch

Surface Element

Drip Pan

Trim Ring

Element Receptacle

Oven Vent

Oven Light

Capillary Tube

Broil Element

Terminal Block

Insulation

Bake Element

Oven Gasket

Contents

How They Work

Electric ranges come in a variety of styles—free-standing, slide-in, double-oven, cooktop, or wall oven—but all operate in much the same way and use similar components. The illustration at left shows a freestanding, four-burner range that contains most of the features found in a typical model. An electric range operates on a 240/120-volt circuit—240 volts for the heating elements and 120 volts for the clock, light, and other accessories. The heating elements are activated by electrical switches that regulate the current reaching the elements to control burner heat. A thermostat senses and regulates oven temperature.

Troubleshooting

Problem	Solution
• **Nothing works**	Check the fuses and the circuit breaker • Test the power cord **44** •
• **No elements heat, or they heat only partially**	Check the fuses and the circuit breaker • Test the power cord **44** • Check connections at terminal block **44** •
• **One surface element does not heat**	Reposition the element **47** • Test the element **49** • Check the element receptacle **47** • Test the burner switch **50** •
• **One surface element provides only high heat**	Replace the burner switch **50** •
• **Oven doesn't heat**	Test the oven element **51** • Test the temperature control **54** • Check the oven selector switch **55** •
• **Oven doesn't hold the set temperature**	Check the capillary tube **54** • Test the temperature control **54** • Recalibrate the temperature control **54** •
• **Oven produces condensation**	Clean the vent; wash the duct **52** •
• **Self-cleaning oven doesn't clean**	Reclose and relock the door • Test the oven element **51** • Test the temperature control **54** • Test the oven selector switch **55** • Door lock broken; call for service • Smoke eliminator faulty; call for service •
• **Oven light out**	Replace the bulb **55** • Test the light switch **56** •
• **Oven door doesn't close properly**	Adjust the door **56** • Adjust or replace the springs **56** • Replace the gasket **57, 58** •

Before You Start

Although an electric range seems a tangle of wires and switches, repairs are based on simple deductive reasoning. Often, solving a problem is as simple as tightening wiring connections.

JUST FOLLOW YOUR NOSE

Since many malfunctions are caused by loose connections and burned wires; always check for these first. Clues to a loose connection are a metallic odor or a soft hissing or buzzing sound. A sharp odor of burning plastic indicates overheating in a switch or terminal block.

Repairing or replacing the burner and oven elements, and adjusting and repairing temperature control mechanisms, are easy tasks that you can do yourself. Fixing poorly seated doors and gaskets, which cause oven temperature to fluctuate, can be done in less than an hour.

Keeping your range clean is the most effective way to avoid breakdowns, but be careful not to get cleaning liquids inside the range, where they can short out circuits. Don't use foil to line the drip pans under the burners or the oven element; it can short the electrical connections. Using the burners without drip pans can also harm the wiring. Never wash the gasket of a self-cleaning oven.

Before You Start Tips:

⋯⋰ Before reconnecting the power to the range after a repair, make sure no uninsulated wires or terminals touch the cabinet, and that the wiring is away from sharp edges

TOOLS

Phillips screwdriver
Standard screwdriver
Nut driver
Multitester
Multipurpose electrical tool
Long-nose pliers
Adjustable wrench
Plastic scraper

MATERIALS

Porcelain connectors
Crimp-on terminals
Steel wool

SAFETY FIRST

Before starting repairs, unplug the range or turn off the power at the service panel. The range draws power through two separate fuses or circuit breakers.

Self-cleaning ovens use high temperatures—about 900°F—to burn off residue. A special locking mechanism prevents the oven door from being opened during the cleaning cycle. Call for service immediately if the mechanism is faulty.

240V Power Cords

1. GAINING ACCESS TO THE TERMINAL BLOCK

The power cords of 240-volt appliances are connected to the machine's internal wiring at a terminal block. If the appliance fails, or if a heating element doesn't heat, check the terminal block and power cord.

• Unplug the range and remove the cover plate where the cord enters the back of the machine.

• Disconnect the three power cord wires from the terminal block *(right)*.

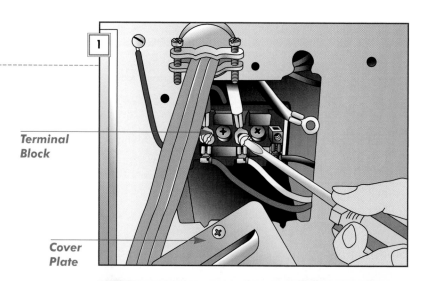

Terminal Block

Cover Plate

2. TESTING THE POWER CORD

• Set a multitester to RX1 and clip a probe to one of the prongs on the power cord plug.

• Touch the other probe to the terminals of the three wires in turn. There should be continuity with one of the terminals *(right)*.

• Repeat this test with the other two prongs. Each should have continuity with a different terminal *(page 18)*.

• Replace the cord if it fails any test *(next step)*.

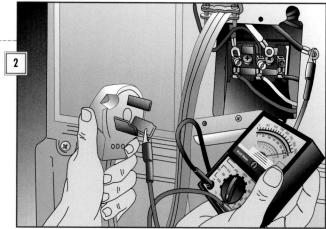

3. REPLACING THE POWER CORD

• If your power cord has four wires, remove the screw securing the ground wire to the cabinet.

• Loosen the metal strain relief clamp, then pull the power cord through the back of the cabinet *(right)*.

• Feed the new power cord through the back of the cabinet and wire it to the terminal block, matching wire colors.

Strain Relief

Ground Wire Screw

Power Cord

Terminal Block

WARNING

Access to the Controls

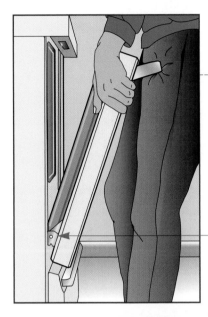

Hinge

REMOVING THE OVEN DOOR FROM A FREESTANDING RANGE

Many oven doors simply slide off their hinges. The door hinges of a self-cleaning oven must be unscrewed from the range.

• Open the door to its first stop.

• Grip each side, maintaining the door's angle, and pull the door straight off its hinges *(left)*.

• For safety, place a piece of scrap wood between the hinge and oven housing, and close the hinge arms against the wood.

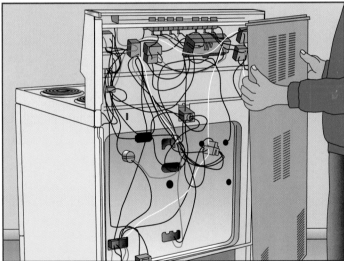

REMOVING THE BACK OF A FREESTANDING RANGE

• Unplug the range or turn off power at the service panel.

• Pull the range away from the wall.

• Support the back panel with a free hand or knee, then remove the screws from around the panel's edges to expose the wiring and controls.

• Some ranges have a single panel *(left);* others have a lower panel covering the terminal block where the power cord is attached, and one or more upper panels covering the wires and controls.

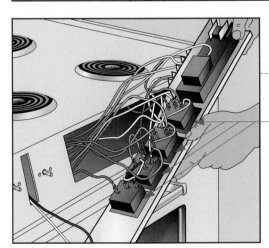

Burner Switches

DETACHING FRONT-MOUNTED CONTROLS

• Disconnect power to the range.

• Remove the screws at each end of the control panel. The panel may also be held in place by a spring clip.

• Tilt the panel forward to expose the controls and wiring.

ACESS TO THE CONTROLS
OF A BUILT-IN RANGE

Often you can work on the controls of a built-in range without moving it.

- Disconnect the power to the range.

- Spread a towel to protect the cooktop.

- Remove the screws from each end cap of the control panel.

- Tilt the panel forward and rest it on the towel.

- Unscrew the rear cover to expose the switches and controls *(right)*.

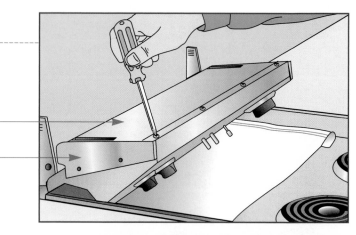

Rear Cover

End Cap

REMOVING THE CONTROL PANEL
FROM A WALL OVEN

- Disconnect the power, then remove the screws at each end of the control panel. You may have to remove the screws from the inside of a recessed wall oven.

- Ease the panel forward *(right),* and rest it on the door or an an oven rack.

- If the control panel is hinged at the bottom, simply open it toward you. In a double wall oven, if the upper oven has a built-in range hood above, release the latch pin under the hood to free the panel.

WHAT'S BENEATH THAT SMOOTH TOP?

The heating elements on smooth-top ranges are mounted beneath a glass surface. Most smoothtops use electric heating elements mounted in ceramic plates. Each element has a thermostat that prevents it from overheating and cracking the glass surface. Unfortunately, the first sign of a bad thermostat is often when the glass cracks from overheating. Perform continuity tests *(page 18)* to locate the faulty thermostat.

To gain access to the heating elements, first remove the control panel. Take out the screws inside the control panel and in the oven compartment securing the cooking surface. Lift the cooktop slightly, then disconnect the wires that prevent it from being lifted off. Service the elements as shown on pages 47-50.

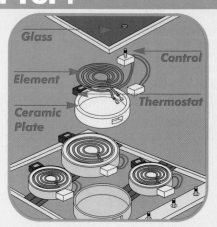

Glass

Control

Element

Thermostat

Ceramic Plate

Plug-In Burners

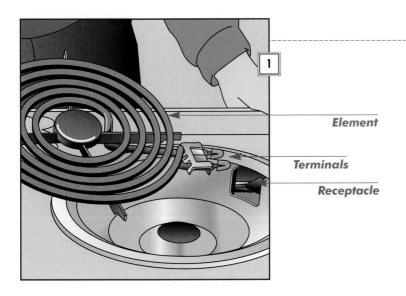

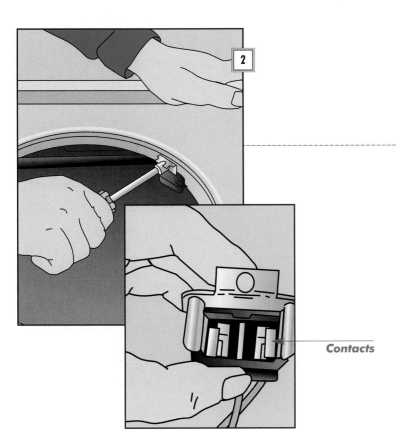

Element

Terminals

Receptacle

Contacts

1. CHECKING A PLUG-IN ELEMENT

Most ranges have sheathed coil elements that plug into receptacles within the burner openings. Test both the element and the receptacle if an element doesn't heat.

• Disconnect power to the range, grasp the element, and reseat its terminals securely in the receptacle. If the problem persists, raise the edge of the element about 1 inch and pull it out *(left)*.

• Inspect the coils for burns or holes. Replace the element if it is damaged. Buff corroded terminals with fine steel wool.

• Test the element by plugging it into a working receptacle. Restore power. Replace the element if it still does not heat.

• If the element heats in another receptacle, check the receptacle from which you unplugged the element *(below)*.

2. CHECKING AN ELEMENT RECEPTACLE

• Lift out the drip pan and its chrome ring. On some ranges you may need to remove screws securing the cooktop and prop the cooktop up to work on the receptacles.

• Disconnect power to the range, then unscrew the receptacle with a screwdriver or nut driver *(left)*.

• Pull out the receptacle (don't strain the wiring) and examine the contacts *(inset)*. Splice on a new receptacle *(page 48)* if the contacts are bent, burned, or oxidized.

• Restore power and test.

3. CLIP TESTER PROBE TO BURNER SWITCH

- Disconnect power to the range.

- Trace the wires from the receptacle to the two corresponding terminals on the burner switch (usually marked H1 and H2).

- Set a multitester to RX1 and clip one probe to one of the corresponding terminals on the burner switch.

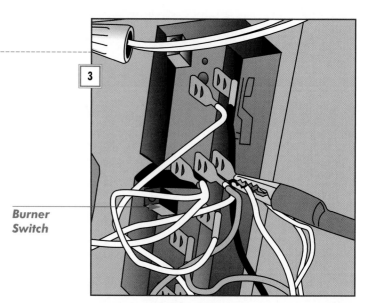

Burner Switch

4. TESTING THE RECEPTACLE CONTACT

- Touch the other probe to each of the receptacle contacts in turn *(right)*. Only one contact should show continuity *(page 18)*.

- Repeat with a probe clipped on the second switch terminal. The other receptacle contact should show continuity. Replace the receptacle if it fails either test.

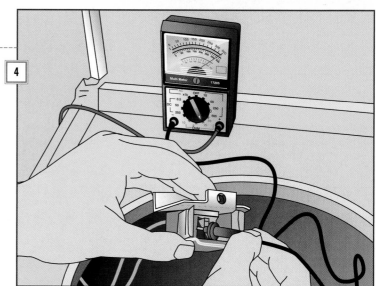

5. REPLACING THE RECEPTACLE

- Cut the wires to the receptacle *(right)*.

- Strip about 1/4 inch of insulation from the wires. Splice the leads from a new receptacle in place using a porcelain wire cap *(inset)*.

- On some models the receptacles' wiring connections are covered by a card-like insulator. To remove the insulator, snap off the clip with a screwdriver, then unscrew the wires. Screw the wires to the new receptacle and clip on the insulator.

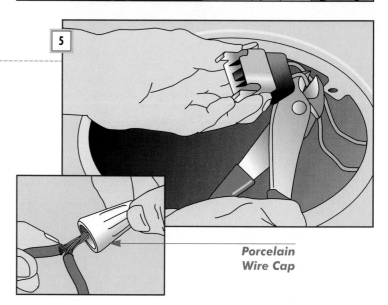

Porcelain Wire Cap

Wired Burners

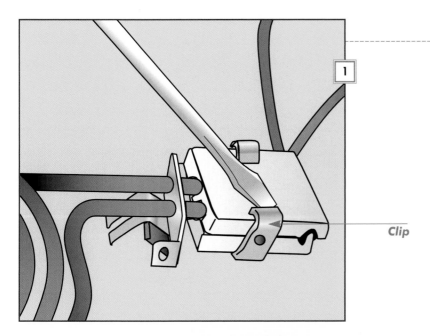

1

Clip

1. DISCONNECTING A WIRED ELEMENT

The burners on some ranges are connected directly to the burner switch wires. The connection is protected by a glass or ceramic block.

• Turn off the power. Remove the drip pan and unscrew the element and block from the range.

• With a screwdriver, pry off the clips joining the two halves of the block *(left)*.

• Tighten loose connections or replace the element if the terminals are burned or corroded.

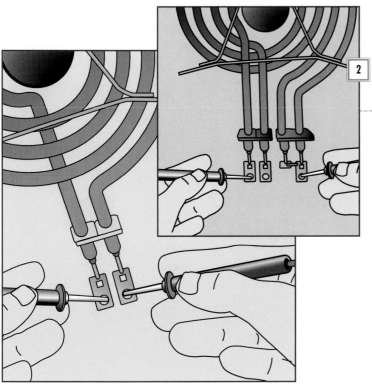

2

2. TESTING THE TERMINALS

• Label and detach the wires from the element.

• Set a multitester to RX1 and touch one probe to each terminal *(left).* The meter should show continuity *(page 18)*. If the element has several terminals *(inset),* half of them are joined to form a common terminal. Touch one probe to this terminal and the other to each terminal in turn. The multitester should show continuity.

• Test for a ground by clipping one probe to the sheathing and touching each terminal; the tester needle should not move *(page 18)*.

• Replace the element if it fails any test.

Burner Switches

1. TESTING SWITCH CONTINUITY

Power-supply wires are attached to terminals L1 and L2, or N1 and N2. Wires leading to the element are marked H1 and H2 (or are simply numbered).

• Disconnect the power and gain access to the range controls and switches *(page 45)*.

• Disconnect one wire from each pair of terminals corresponding to the suspect burner. Turn the switch for the suspect burner to the ON position.

• Set a multitester to RX1 and test for continuity between L1 and H1 and L2 and H2 *(page 18)*. Replace the switch if there is no continuity.

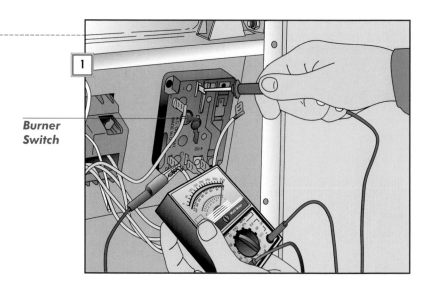

Burner Switch

2. REMOVING THE SWITCH

• Pull off the control knob.

• Remove the screws holding the switch to the control panel *(right)*. If a glass panel covers the screws, check for clips or trim pieces at the top and sides of the panel. Unclip or unscrew, as needed, then lift out the panel.

• Pull the switch out from the back.

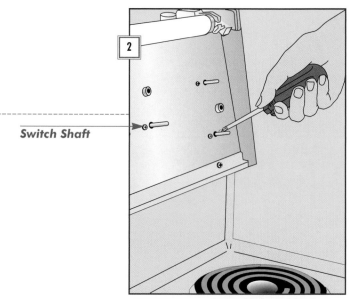

Switch Shaft

3. REPLACING THE SWITCH

• Leave the wires connected to the faulty switch until you have a replacement part.

• One at a time, transfer the wires from the old switch to the corresponding location on the new switch *(right)*.

• Screw the new switch to the control panel and replace the control panel cover, trim, and control knobs.

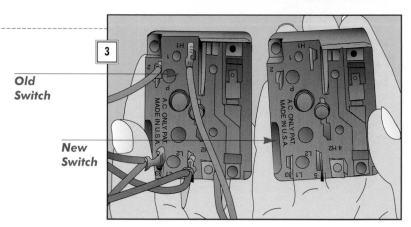

Old Switch

New Switch

Oven Elements

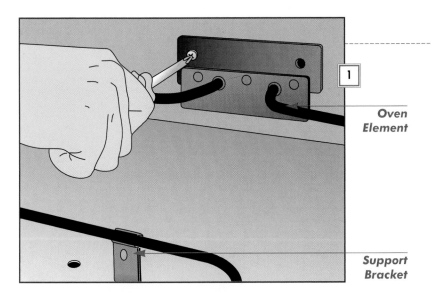

Oven Element

Support Bracket

1. UNMOUNTING OVEN ELEMENTS

BAKE and BROIL elements are both tested and replaced in the same way. Remove the oven door for easy access *(page 45)*.

• Take out the screws or nuts that fasten the element to the back of the oven *(left)*.

• If the capillary tube *(page 52)* is in the way of the element, unclip it from its support, being careful not to bend it.

• Unscrew any support brackets. Pull the element forward to expose wiring.

⚠ **CAUTION: In self-cleaning ovens, the capillary tube contains a caustic fluid. Wear rubber gloves and handle it gently.**

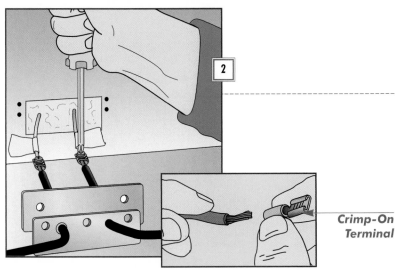

Crimp-On Terminal

2. DISCONNECTING THE ELEMENT

• Label wires with masking tape to record their positions, then unscrew or unplug them from the element terminals *(left)*. Avoid bending the terminals, and don't let the wires fall back through the opening.

• Remove the element from the oven.

• Check the wire terminals for burns; if damaged, cut them off and replace them with crimp-on terminals *(inset)*.

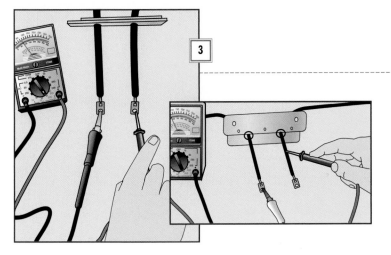

3. TESTING AND REPLACING THE ELEMENT

• Set a multitester to RX1 and touch probes to each of the terminals *(left)*. The meter should show continuity *(page 18)*.

• Test for ground with one probe on a terminal and the other on the element's sheath *(inset)*. The needle shouldn't move.

• Replace a faulty element with one of equal wattage.

Calibrating Oven Temperature

1. CLEANING THE VENT AND DUCT

The oven vent helps control air circulation and temperature by expelling hot air through a duct.

• Disconnect power to the range and remove the burner and drip pan.

• Unscrew the duct if necessary and lift it away to expose the vent *(right)*. Clean the vent and wash the duct.

• Be sure the duct opening lines up with the hole in the drip pan.

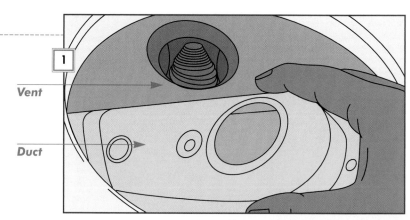

Vent

Duct

2. CHECKING THE CAPILLARY TUBE

A capillary tube senses the temperature and activates the oven temperature control.

• If the capillary tube touches the oven wall, reposition it in its support clips *(right)*.

• If it is damaged, replace the tube and temperature control switch.

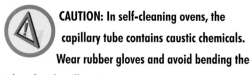 **CAUTION: In self-cleaning ovens, the capillary tube contains caustic chemicals. Wear rubber gloves and avoid bending the tube when handling it.**

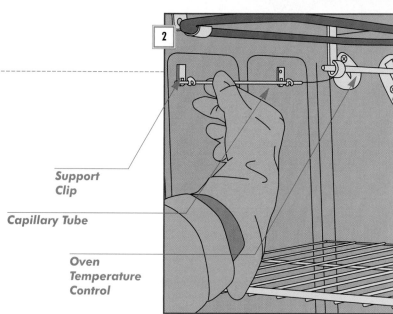

Support Clip

Capillary Tube

Oven Temperature Control

3. TESTING OVEN TEMPERATURE

• Place a thermometer in the oven and set the temperature control at 350°F *(right)*.

• Wait 20 minutes. Check the temperature four times over the next 40 minutes and calculate the average.

• If the result is off by less than 25°F, the control is normal; by 25°F to 50°F, recalibrate the temperature control *(Step 4)*. Replace the control if the temperature is off by 50°F or more.

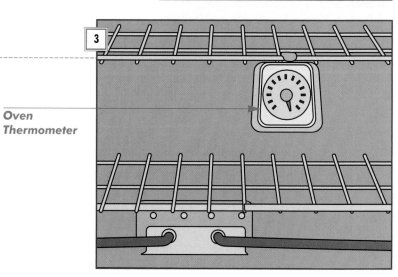

Oven Thermometer

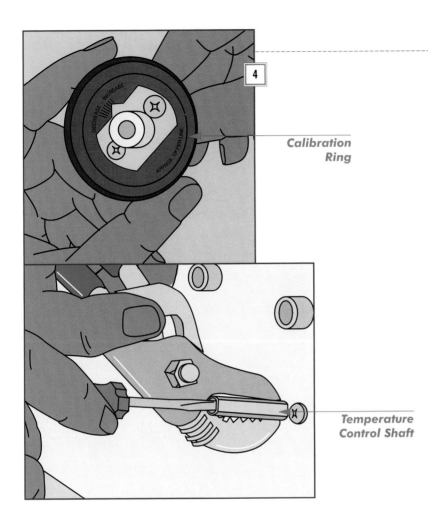

Calibration Ring

Temperature Control Shaft

4. CALIBRATING OVEN TEMPERATURE

Some ovens have a calibration ring on the back of the temperature-control knob. On others, temperature adjustments are made with a screw inside the control-knob shaft.

• Pull the control knob off the oven temperature control.

• If the knob has a ring with marks indicating DECREASE and INCREASE *(top left)*, loosen the screws and turn the knob to move the ring in the appropriate direction. Retighten the screws.

• Where there is no ring *(bottom left)*, hold the control shaft still with adjustable pliers. Insert a thin screwdriver, and chip away the factory seal. Then turn the inside screw clockwise to raise the temperature and counterclockwise to lower it. A 1/8 turn adjusts the setting about 25°F.

• Retest the oven temperature.

• Replace the temperature control if the calibration does not work *(page 54)*.

Ron's TRADE SECRETS

CHECKING THE THERMOSTAT BULB

If an oven isn't holding its temperature, the first thing I do is examine the thermostat bulb. The bulb is often clipped to the back of the oven, and can easily be bumped out of position by cookware or when an oven rack is moved to a different location. This will prevent the oven from registering the correct heat setting. The fix is simple: Resecure the bulb to its clip, as shown here. I always take a minute to examine the capillary tube beneath the bulb. If it is bent or crimped (or if the bulb is damaged), I replace the entire thermostat. If everything looks to be in good shape, I test the oven temperature using the methods described above.

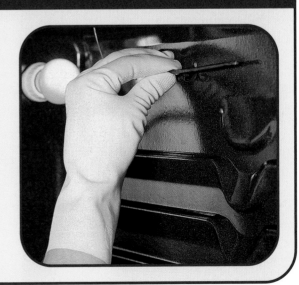

Oven Temperature Controls

1. TESTING THE TEMPERATURE CONTROL

• Turn off the power and open the control panel *(page 45)*.

• If any of the terminals appear discolored or burned, replace the temperature control.

• If the control has more than two terminals, identify which terminals to test using the diagram located on the rear panel, or inside the storage drawer or control panel.

• Set a multitester to RX1. Disconnect one wire from the terminals being tested and clip on tester probes *(right)*.

• Set the oven temperature dial to 300°F.

• Replace the control if the meter doesn't indicate continuity *(page 18)*.

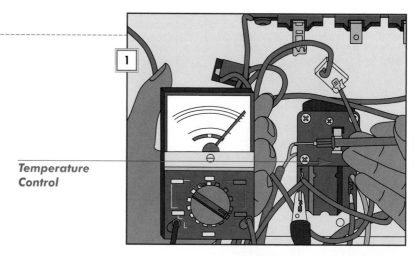

Temperature Control

2. REMOVING THE CAPILLARY TUBE

• Unclip the tube from its supports in the oven *(page 52)* and push it through the rear wall *(right)*. If needed, loosen the screw that secures the baffle and slide it aside.

• From the back of the range, pull the tube completely out of the oven *(inset)*.

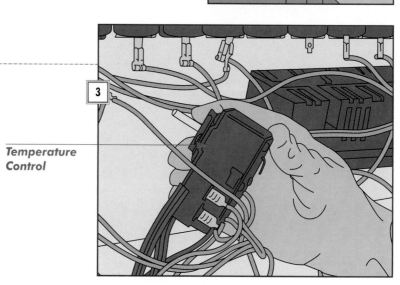

Baffle
Capillary Tube

3. REMOVING AND REPLACING THE TEMPERATURE CONTROL

• Unscrew the two temperature control screws in the front and remove the control from the back of the range *(right)*.

• Label and disconnect the wires; replace any burned wire connectors *(page 51)*.

• Connect the new control and screw it in place.

• Gently push the capillary tube through the back and into the oven, then clip it to its supports.

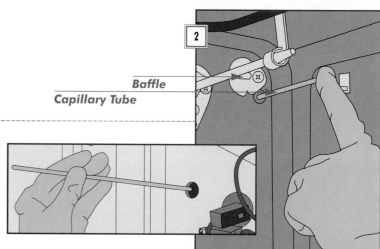

Temperature Control

Oven Selector Switch

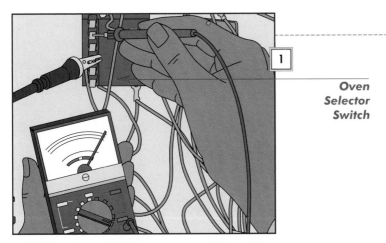

Oven Selector Switch

1. TESTING SWITCH CONTINUITY

The oven selector switch controls the BAKE, BROIL, TIMED BAKE, and CLEAN functions. Replace the switch if any of the terminals appears burned *(next step)*.

• Disconnect power to the range and open the control panel *(page 45)*.

• Set a multitester to RX1 *(page 18)*. Disconnect one wire from each pair of terminals being tested and check for continuity at each position *(left)*.

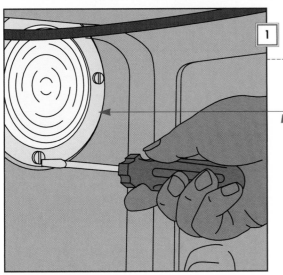

Oven Selector Switch

2. REPLACING THE SWITCH

• Remove the screws from the front of the control panel and pull the switch out of the back *(left)*.

• Label and disconnect the wires; replace any burned wire connectors *(page 51)*.

• Reconnect the wires to the terminals on the new switch and screw the switch securely to the control panel to ensure proper grounding.

Oven Light

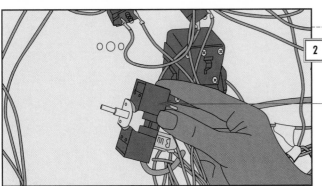

Bulb Shield

1. REPLACING THE OVEN LIGHT BULB

• Pull down the wire bulb protector or unscrew the glass shield *(left)*.

• Remove the bulb and install a standard 40-watt bulb as a test; if it works, replace it with an appliance bulb of the same size and wattage as the original.

• If the bulb doesn't light, check the switch *(next page)*.

2. TESTING OVEN LIGHT SWITCHES

If the oven has a door-operated light switch, follow the same test procedures as for a refrigerator light switch *(page 17)*.

● Disconnect power to the range and open the control panel *(page 45)*.

● Set a multitester to RX1. Remove one wire from the switch and connect one probe to each terminal *(right)*.

● Flip the switch on; the multitester should show continuity *(page 18)*.

● Replace the switch if it fails the test. Push the switch out through the front of the range; connect the wires to a new switch.

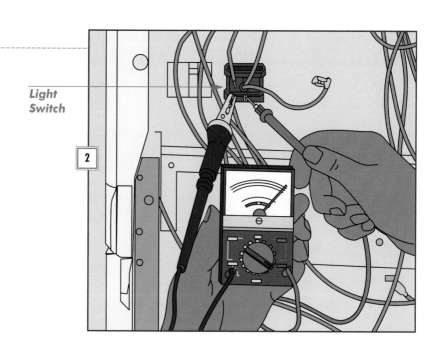

Light Switch

2

Oven Door

1. ADJUSTING THE FIT

● Open the door and loosen the screws securing the inner panel. Hold the door at the top. Gently twist it from side to side to straighten it and seat it securely on its hinges *(right)*. Shift the door only slightly if it has a glass front or window.

● Partially tighten the door screws; do not overtighten—the porcelain could chip.

● To check the seal, insert a dollar bill between the seal and the top corners of the oven. The seal should tightly grip the bill.

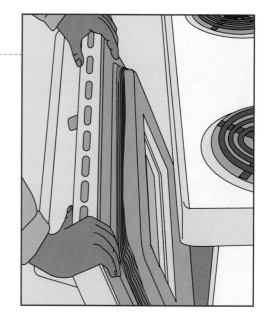

2. REPLACING CABINET-MOUNTED SPRINGS

● Remove the door *(page 45)*.

● Pull out the lower storage drawer. If you can't find the springs, remove the side panel or call for service.

● Wearing safety goggles, unhook and replace the springs. Replace both springs even if only one is broken.

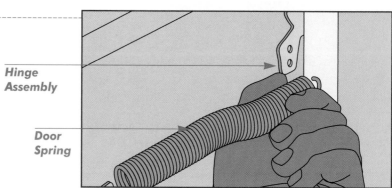

Hinge Assembly

Door Spring

Hook-Style Oven Gaskets

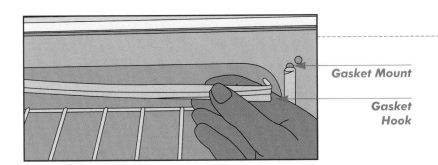

Gasket Mount

Gasket Hook

REPLACING A SIMPLE OVEN GASKET

The gasket on many ovens is a simple rubber channel clipped to the cabinet.

• Visually inspect the gasket for damage.

• Disengage any damaged sections by hand and hook new ones in place *(left)*.

Clamped Oven Gaskets

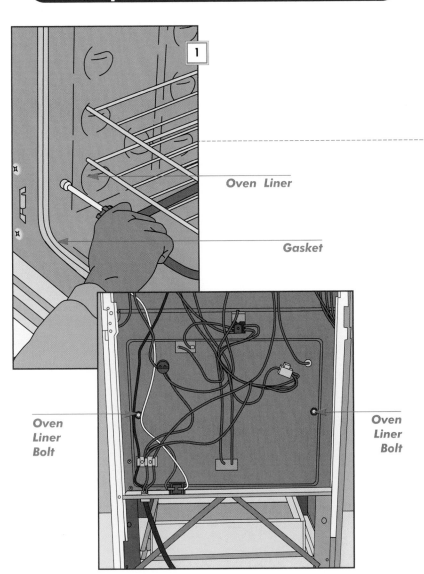

Oven Liner

Gasket

Oven Liner Bolt

Oven Liner Bolt

1. GAINING ACCESS TO A CLAMPED GASKET

Cabinet-mounted gaskets are clamped between the oven liner and range cabinet.

• Remove the door *(page 45)*.

• Check around the front edge of the oven for screws or clips holding the oven liner in place. Remove the screws *(top left)*.

• If there are no screws inside the oven, disconnect power to the range and pull it away from the wall. Locate the oven liner bolts that protrude from the back of the range and loosen them about a 1/4 inch *(bottom left)*.

• If you cannot identify the bolts, call for service.

2. REMOVING AND REPLACING A CLAMPED GASKET

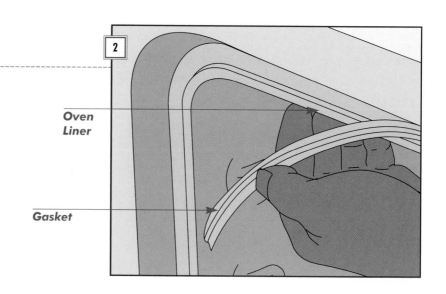

- Partially pull out the oven liner by rocking it back and forth.

- Disengage the gasket from between the liner and cabinet *(right)*.

- Position the lip of a new gasket behind the rim of the oven liner.

- Push the oven liner back into place and replace the screws or tighten the bolts in the back of the oven.

Oven Liner

Gasket

Door-Mounted Gaskets

1. SEPARATING THE DOOR PANELS

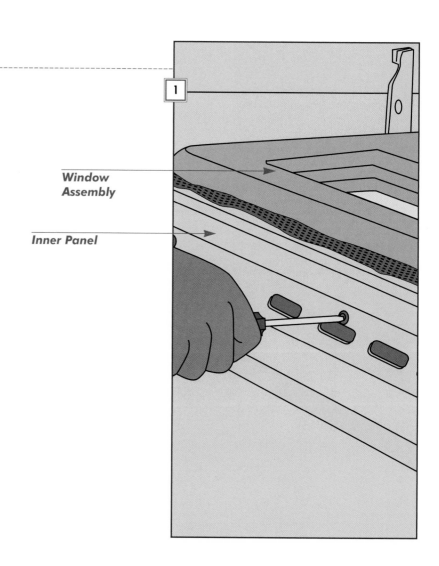

On self-cleaning ovens, the gasket is held between the panels of the oven door and can be replaced only by disassembling the door.

- Unscrew the door-hinge arms and take off the door.

- Remove the screws on the inner panel and along the outer edges of the door *(right)*. On some models, you may have to remove the door handle as well.

- If there are tabs on the outer panel that fit into slots in the inner panel, gently pry them apart with a screwdriver.

- Starting at the top, lift off the inner panel and window assembly.

Window Assembly

Inner Panel

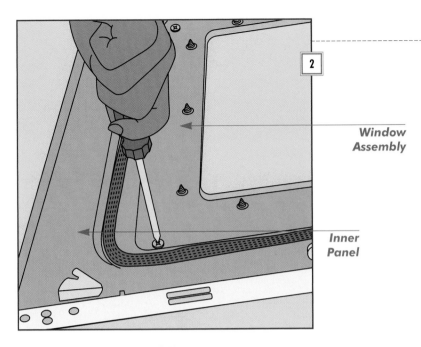

Window
Assembly

Inner
Panel

2. REMOVING A DOOR WINDOW

• Remove the screws that hold the window assembly to the panel *(left)*. To reach the window on some models you must first remove a metal window shield and a layer of insulation.

• Lift off the window assembly to reveal the gasket attachment.

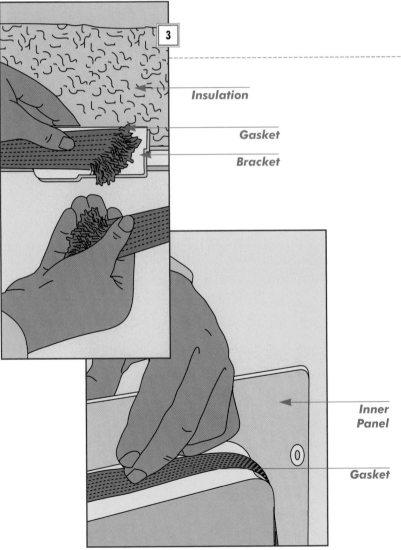

Insulation

Gasket

Bracket

Inner
Panel

Gasket

3. REPLACING THE GASKET

• Locate the gasket between the window assembly and the inner door panel.

• Unscrew the clips (if any) that hold the gasket in place and slip out the gasket.

• Position the new gasket with the small bead, or edge, against the panel edge, then hook the two gasket ends to the bracket *(top left)*.

• Loosely screw the window assembly and the door panel together.

• To adjust a loose-fitting gasket, use the tip of a plastic scraper to wedge the excess between the door panel and the window assembly *(bottom left)*.

• Starting at the top of the door, push the gasket in with the scraper, tightening the screws as you go.

• Screw the inner and outer door panels back together and reinstall the door on the oven, tightening the screws securely.

• If necessary, adjust the fit of the oven door *(page 56)*.

FIX IT... Gas Ranges

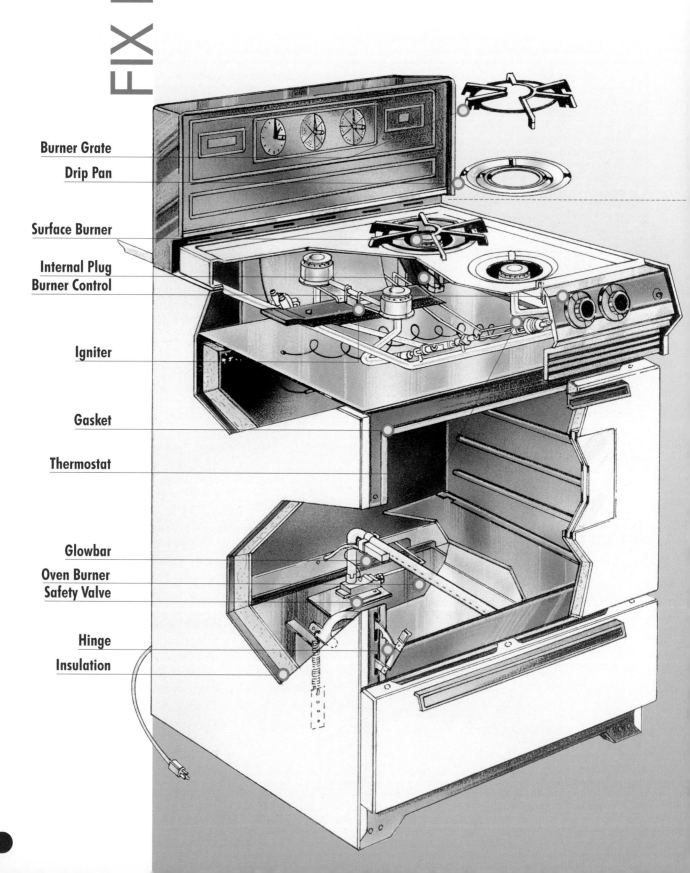

Burner Grate

Drip Pan

Surface Burner

Internal Plug
Burner Control

Igniter

Gasket

Thermostat

Glowbar

Oven Burner
Safety Valve

Hinge

Insulation

Chapter 3

Contents

How They Work

Some gas ranges have a pilot flame that ignites the fuel flowing to the surface and oven burners when the control knobs are opened. Others have spark igniters or an electrically-heated coil—called a glow-bar—to ignite the gas. The igniters are wired to an ignition module on the back of the range; it produces the high voltage required for sparking. On both types, air shutters control the amount of air mixed with the gas flowing to the burners. A thermostat regulates oven temperature by turning the gas supply on and off.

Troubleshooting

Problem	Solution

Gas odor

Ventilate room; relight pilot **65**

Gas odor with all pilots lit or with electric ignition

Turn off burner controls
Turn off gas to range; call gas company **65**

Clock, lights, igniters don't work

Check internal range plug **64**

Surface burner doesn't light

Check that range is plugged in (ranges with electric igniters)
Relight pilot **65** Adjust pilot **66** Clear burner ports **66**
Clean surface burner **66** Adjust air shutter **67**
Inspect igniter and ignition module (ranges with electric igniters) **68**

Surface burner pilot doesn't stay lit

Clear pilot opening **65** Adjust pilot **66** Adjust air shutter **67**

Surface burner flame uneven or low

Clear burner ports; clean surface burner **66** Adjust air shutter **67**

Surface burner flame too high, noisy, or blowing

Adjust air shutter to reduce air **67**

Surface burner flame yellow, sooty

Adjust air shutter to increase air **67**

Oven burner doesn't light

Check that range is plugged in (ranges with electric igniters)
Relight pilot **69** Adjust pilot **66** Clear oven burner ports **70**
Inspect igniter and ignition module (ovens with electric igniters) **72**
Test fuse (ovens with glowbar igniters) **73** Test glowbar **73**

Oven burner pilot doesn't stay lit

Clear pilot opening **65** Adjust pilot **69**

Oven doesn't operate at set temperature or bakes unevenly

Check door **56** Check gasket **57** Clear oven burner ports **70**
Thermostat or capillary tube faulty; call for service

Self-cleaning oven doesn't clean

Reset control; consult owner's manual
Check door **56** Check gasket **57**
Thermostat, selector switch, or lock faulty; call for service

Before You Start

The two basic types of gas ranges differ mainly in the way they ignite gas. One is the traditional pilot flame; the other, an electronic igniter.

A SIMPLE APPLIANCE

A gas range has few moving parts, and many older stoves have no electrical components at all. Repairs on the following pages can be adapted to older and newer models alike. For example, in the illustrations regarding adjustments and repairs to oven burners (*pages 69 to 73*), you'll see a couple of different designs for safety valves—all have similar parts and can be worked on in similar ways.

Servicing and reigniting pilot lights, and cleaning and adjusting surface and oven burners, are easy tasks that you can do yourself. Some gas range repairs—such as replacing a gasket or adjusting an oven door—are the same as for electric ranges. If a repair is not listed here, consult the electric range Troubleshooting Guide (*page 42*).

Before You StartTips:

⋯⋗ Repairs that involve the gas-supply line, such as replacing a burner control or a thermostat, carry the risk of creating gas leaks and should be handled by a professional.

⋯⋗ Tests on electronic ignition-system components require a multitester (page 18).

⋯⋗ Make sure you know the location of the gas shutoff valve for your range, as well as for the gas service to your house—and know how to operate them (*page 65*).

TOOLS

Screwdriver

Wrenches

Multitester

Wire cutter

Wire stripper

Crimping tool

MATERIALS

Matches

Clean rag

Sewing needle

Electrical tape

Porcelain wire caps

SAFETY FIRST

If the gas-supply line is a rigid pipe, never move the range—you can cause a gas leak. Have the gas company or a service technician disconnect and move it for you.

Access to the Range

OPENING THE COOKTOP

In gas ranges with electrical components—clocks, lights, spark igniters—an internal harness plug connects the power cord to the internal wiring. Unplug this assembly to disconnect the range *(below)*.

• Remove the burner grates. Grasp the front edge of the cooktop and lift.

• Prop the cooktop open with the support rod located inside the range hood *(right)*.

• On some ranges you can completely remove the cooktop by lifting it up and then pulling it out toward you.

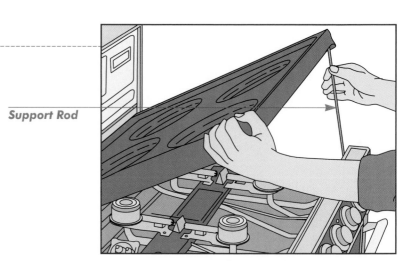

Support Rod

DISCONNECTING THE INTERNAL RANGE PLUG

• Disconnect the plug (usually at the right rear of the range).

• Examine the plug terminals *(right)*.

• Straighten bent terminals, if possible. Label the wires. Splice on a replacement plug using heat-resistant porcelain wire caps if the terminals are burned *(page 48)*.

• Reconnect the plug. If the electrical components don't work, call for service.

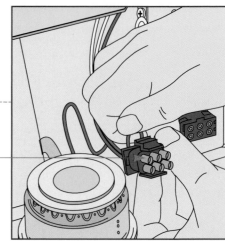

Terminals

REMOVING THE OVEN BOTTOM

• On many ranges you can simply lift out the oven bottom. Others require that you slide forward the small locking tabs at the front or rear of the oven bottom *(right)*.

• Remove any screws holding the oven bottom in place.

• To remove the baffle beneath the oven bottom, take out the wing nuts or screws holding it in place; lift it up and out *(inset)*.

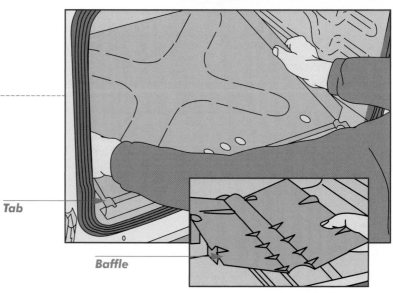

Tab

Baffle

Surface Burner Pilots

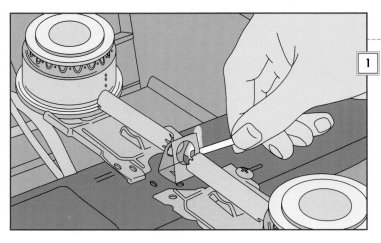

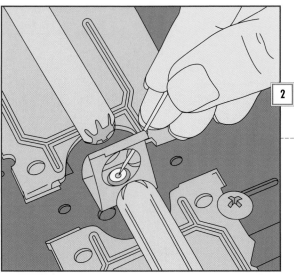

1. RELIGHTING THE PILOT

Pilot lights that blow out often may be set too high or too low, or the pilot-burner may be clogged, hindering gas flow.

- Turn off all of the range controls and prop open the cooktop *(page 64)*.

- Place a lighted match near the opening of the pilot, located midway between the two burners *(left)*.

- Clean or adjust the pilot if it does not stay lit *(next step)*.

 CAUTION: If the pilot light has been out for awhile, or if you detect a gas odor, ventilate the room and call for service.

2. CLEANING THE PILOT

- Remove the metal shield (if necessary) by pressing its tabs on either side.

- Insert a sewing needle in the pilot opening and move it up and down to remove any obstructions *(left)*.

- Clean an oven burner pilot the same way.

TURNING OFF THE GAS SUPPLY

Turn off the gas supply to any appliance that you suspect has a gas leak. If there's a valve on its gas-supply pipe, turn the handle perpendicular to the pipe to shut off the gas *(near right)*. If the appliance supply line doesn't have a shutoff, turn off the gas at the meter in the same way *(far right)*. If the gas odor does not dissipate, leave the house and call the gas company.

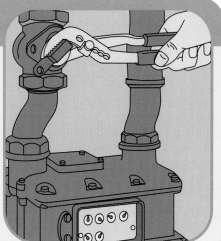

3. ADJUSTING THE PILOT HEIGHT

• Turn the burner controls to OFF and prop open the cooktop *(page 64)*.

• Locate the pilot adjustment screw on the side of the pilot, or on the pilot gas line near the manifold at the front of the range, or behind the burner control knob.

• Turn the screw counterclockwise to increase the size of the pilot *(right)*. The flame should be a sharp, blue cone, 1/4 to 3/8 inch high *(inset)*.

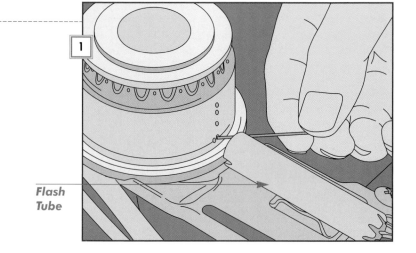

Adjustment Screw

Surface Burners

1. CLEANING BURNER PORTS

If a burner will not light, raise the cooktop *(page 64)* and check that the pilot light or spark igniter is working *(pages 65 and 68)*. A shutter or sleeve on the burner tube controls the air-gas mix *(page 67)*.

• Check that the burner is properly seated. The flash tube must align with the burner ports and the pilot or spark igniter.

• Use a needle to clear the burner ports opposite the flash tube *(right)*.

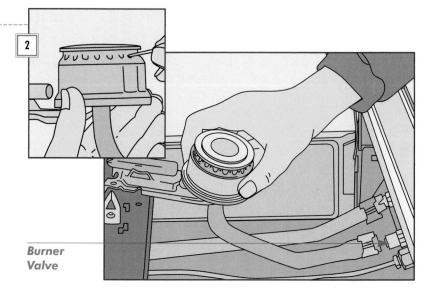

Flash Tube

2. REMOVING A SURFACE BURNER

• Prop open the cooktop *(page 64)*.

• Lift the burner from its support, then pull the burner backward off the burner valve *(right)*. On some older ranges, only the top ring of the burner comes off.

• Wash the burner in hot, soapy water and scrub the portholes with a brush. Clean out the flame openings with a needle *(inset)*. Let the burner dry thoroughly.

• Slip the burner onto its supply valve and rest it securely on its support.

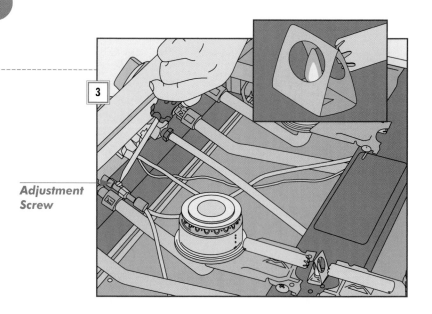

Burner Valve

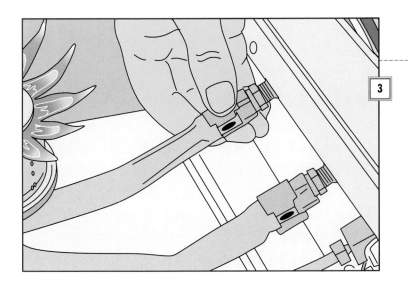

3. ADJUSTING A SURFACE BURNER AIR SHUTTER

• Turn all of the controls to OFF, then raise the cooktop *(page 64)*.

• Locate the air shutter. If it has a setscrew holding it in place, loosen it.

• Turn the burner control on HIGH. Twist or slide the shutter open *(left)* until the flame shows symptoms of excessive air *(below)*. On models with an air-mixing chamber, loosen the retaining screw, slide the plate to adjust the air intake, then retighten the screw.

• Slowly close the shutter until the flame assumes the correct size and color.

• Turn off the burner, tighten the shutter screw, and replace the cooktop.

ANALYZING THE FLAME

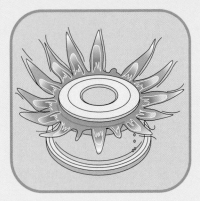

INSUFFICIENT AIR
Not enough air results in a weak flame without sharp blue cones. The flame may be red, yellow, or yellow-tipped and may leave soot deposits on pots and pans.

EXCESSIVE AIR
Too much air causes an unsteady, blowing flame and may prevent burner ignition. If the burner lights, the flame may burn unevenly and could be noisy.

CORRECT AIR ADJUSTMENT
A properly adjusted flame burns steadily, quietly, and uniformly. The cones are blue in color and sharply defined—about 1/2 to 3/4 inch in length.

Surface Burner Spark Igniters

CHECKING THE IGNITER

Observe the igniter as you turn the control to light the burner. If it doesn't spark, turn on the other burner served by the same igniter. If the igniter sparks, the first burner control is faulty; call for service. If neither control activates the spark, the igniter or ignition control module needs replacement.

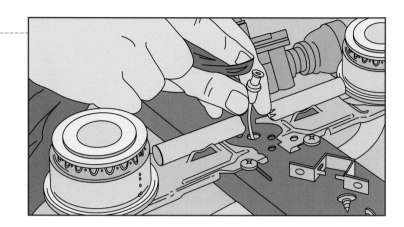

- Unscrew the bracket and inspect the igniter. Wipe the igniter with a rag *(right)*, then recheck it as described above.

- Replace the igniter if it is cracked or burned *(below)*.

INSTALLING A NEW IGNITER

- Trace the igniter cable to its terminal on the ignition control module at the back of the range *(right)*. Unscrew the cover, and disconnect the cable *(inset)*.

- Clip the old igniter off its cable. Tape the end of a new cable to the cut end of the old one. Pull on the old cable to thread the new cable to the ignition control module.

- Connect the end of the new igniter cable to its terminal on the ignition control module.

- Seat the new igniter firmly on the burner support and refasten the bracket.

Burner Support

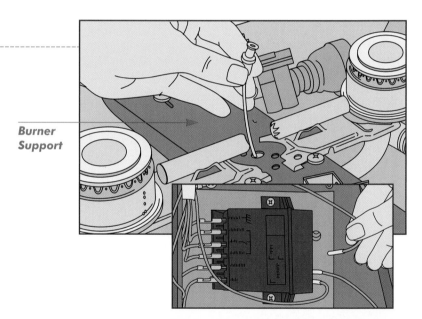

REPLACING THE IGNITION CONTROL MODULE

- Locate the ignition control module at the back of the range.

- Label and disconnect all of the wires leading to the module.

- Unscrew and remove the old ignition control module *(right)*.

- Screw a new module into place and reconnect the wires.

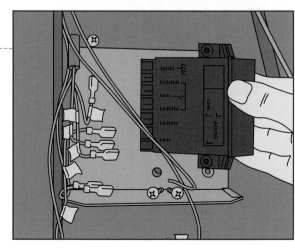

Oven Pilot

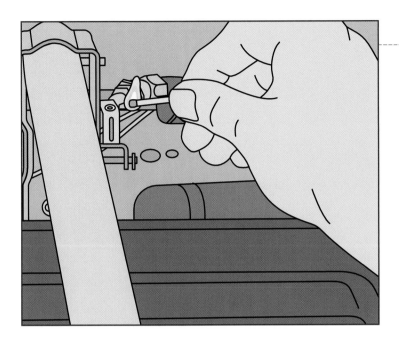

RELIGHTING THE OVEN PILOT

• Turn all of the range controls to OFF.

• If the range is equipped with a broiler drawer, open it and wait for any gas to dissipate. Otherwise, open the oven door and remove the oven bottom and baffle *(page 64)*.

• Place a lighted match near the tip of the pilot on the burner assembly *(left)*. On some older ranges you must hold down a button on the side of the oven or on the thermostat as you light the pilot.

• Turn on the oven. If the burner does not light in a minute or two, adjust the pilot *(below)*. Otherwise, turn off the oven and reinstall the oven bottom and baffle.

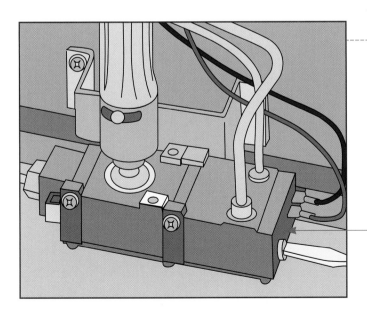

Safety
Valve

ADJUSTING A PILOT (SAFETY VALVE)

• Turn all of the range controls to OFF.

• Open the oven door and remove the oven bottom and baffle, if you haven't already done so *(page 64)*.

• On the safety valve at the rear of the oven, find a slotted screw near the pilot gas line. If there is no adjustment screw, refer to page 70.

• If the pilot will not light, turn the screw slightly counterclockwise *(left)*, then light the pilot. Continue to turn the screw in tiny increments until the pilot has a blue flame.

• If the burner doesn't light within a couple of minutes, the safety valve needs replacing. Call for service.

ADJUSTING THE PILOT (THERMOSTAT)

• Pull off the oven control knob and locate the oven-pilot adjustment screw (sometimes marked P or PILOT) on the front of the thermostat *(right)*. Sometimes you'll find more than one adjustment screw for regulating other pilots within the range. If you can't find the screw at the front of the thermostat, raise the cooktop *(page 64),* and inspect the back of the thermostat near the oven-pilot gas line.

• Open the oven door and remove the oven bottom and baffle *(page 64).*

• Adjust the pilot to a tight, blue flame *(inset).* If the screw has no effect, the thermostat may be defective; call for service.

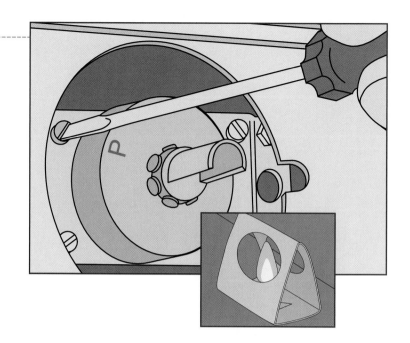

Oven Burner

CLEANING AN OVEN BURNER

Poor baking results or an odor of gas when the oven is turned on may be due to an uneven flame produced by a clogged burner.

• Turn all of the range controls to OFF.

• Open the oven door and remove the oven bottom and baffle *(page 64).*

• Turn on the oven and observe the burner. If the flame is not continuous along the length of the burner, some of its flame ports may be clogged.

• Turn off the oven control and wait for the burner to cool.

• With a needle, clear each port in the burner *(right).*

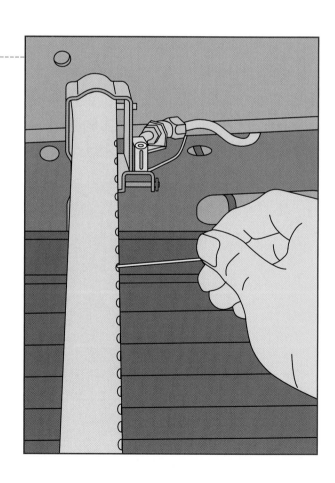

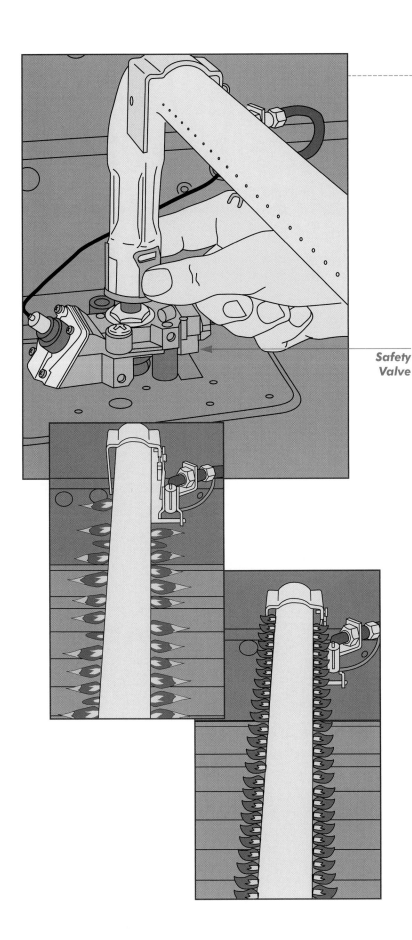

Safety
Valve

ADJUSTING THE OVEN FLAME

• Turn all of the range controls to OFF. Open the oven door and remove the oven bottom and baffle *(page 64)*.

• Turn on the oven and observe the flame. A ragged, hissing flame *(top inset)* shows that there is too much air; a yellow-orange flame indicates too little air.

• If the flame needs adjusting, turn off the oven and let it cool.

• Find the air shutter at the base of the oven burner, just above the safety valve. Some older ranges have an air-mixing chamber with a sliding plate that controls the size of the air opening.

• Loosen the setscrew that locks the shutter or plate, and adjust the opening to increase or decrease the amount of air mixing with the gas *(left)*. Then turn the oven on again.

• Repeat this process until you have adjusted the burner to produce a steady, blue, 1-inch flame with a distinct core about 1/2 inch long *(bottom inset)*.

• When the flame is at its best, turn off the oven and tighten the setscrew.

Oven Spark Igniter

1. CHECKING THE IGNITER

The igniter is connected to the ignition control module.

- Turn all of the range controls to OFF.

- Open the oven door and remove the oven bottom and baffle *(page 64)*.

- Unscrew the igniter from its mounting bracket and inspect it for cracks or other flaws *(right);* replace it if it is damaged.

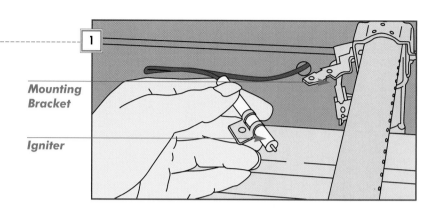

Mounting Bracket

Igniter

2. REPLACING THE IGNITER

- Trace the cable to the ignition module and unscrew the cover.

- Thread in the cable of a new igniter, as described on page 68.

- Seat the new igniter to the bracket with its electrode 1/4 inch from the pilot.

- Connect the cable to the module *(right)*.

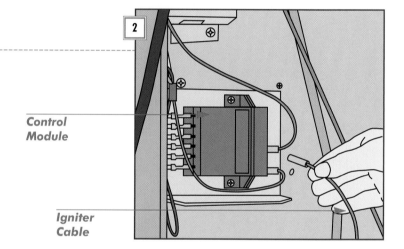

Control Module

Igniter Cable

Ron's TRADE SECRETS

CLEANING BURNERS

I get a lot of calls to repair burners on gas ranges with electronic ignitions that won't ignite. You'd be surprised how often the reason for the breakdown isn't years of wear but a homeowner's good intentions. Many people use steel wool scouring pads to clean up tough stains or spills on a burner. While this may seem the best way to tackle the job, steel wool pads leave behind fibers *(inset)* that short out the spark electrode that ignites the burner. My advice is to clean spills and other residue with a mildly abrasive cleaner and a pot-scrubbing sponge with an abrasive pad on one side. You can also use a plastic wool substitute, available in 8- by 4-inch pads, anyplace you don't want to use steel wool.

Oven Glowbar Igniter

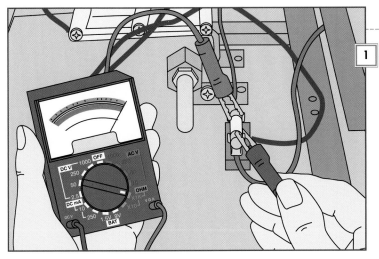

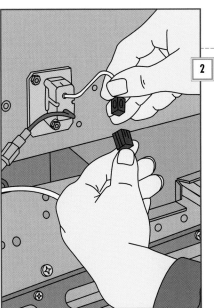

Terminal
Block

Glowbar

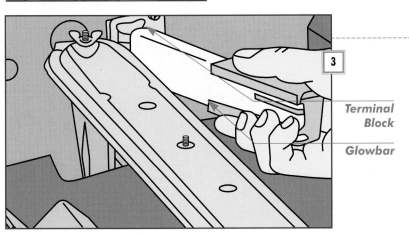

1. TESTING THE FUSE

In some gas ranges, an electrically heated glowbar ignites the oven burner.

• Raise the cooktop and disconnect the internal range plug *(page 64)*.

• Remove oven bottom and baffle *(page 64)*. Find the fuse near the safety valve.

• Set a multitester to RX1 and disconnect one of the wires to the fuse.

• Touch one probe to each end of the fuse *(left)*. The meter should show continuity *(page 18)*; if it doesn't, replace the fuse.

2. CHECKING THE GLOWBAR

• Turn on the thermostat; if the indicator light does not come on, call for service.

• Turn off power to the range.

• Unscrew the cover plate over the glowbar wiring at the back of the range and disconnect the plugs *(left)*, or find the glowbar beneath the oven baffle *(page 64)*.

• Set a multitester to RX1. Touch a probe to each glowbar terminal; replace the glowbar if it does not show continuity.

• If the glowbar has continuity, the safety valve may be faulty; call for service.

3. REPLACING THE GLOWBAR

• Unscrew the glowbar from the burner support bracket and the oven wall.

• Pull the glowbar free of the terminal block *(left)*.

• Mount a new glowbar in the burner support bracket at the back of the range.

• Reconnect its plugs at the back of the range and replace the metal cover.

FIX IT: Dishwashers

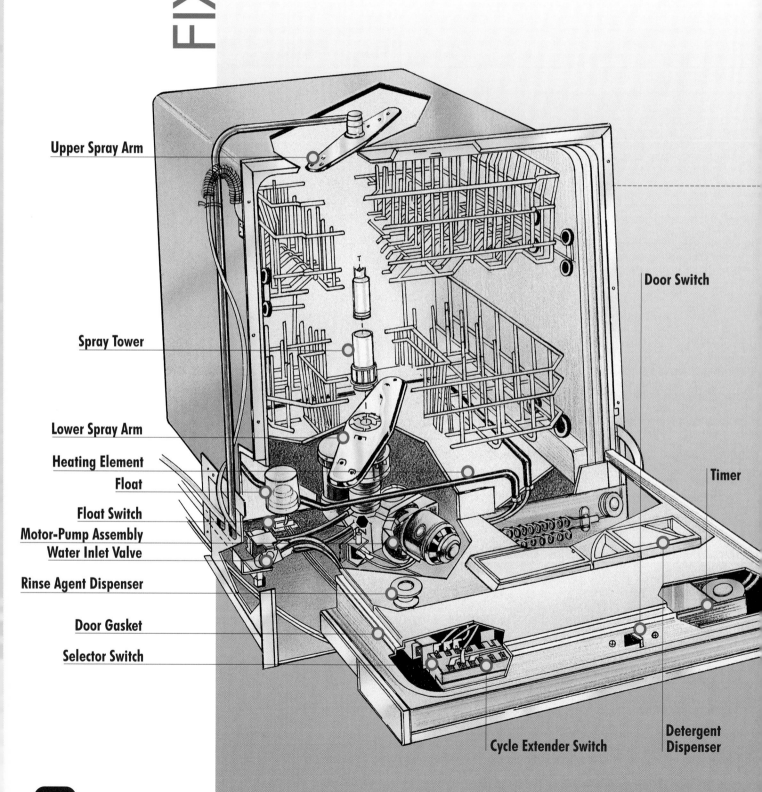

Upper Spray Arm

Spray Tower

Lower Spray Arm

Heating Element

Float

Float Switch

Motor-Pump Assembly

Water Inlet Valve

Rinse Agent Dispenser

Door Gasket

Selector Switch

Door Switch

Timer

Cycle Extender Switch

Detergent Dispenser

Chapter 4

Contents

How They Work

A dishwasher combines water, detergent, and heat to clean dishes. A preset amount of water flows through the water inlet valve into the tub. The detergent dispenser then releases its contents. A heating element warms the detergent-and-water solution, then a pump channels it to the spray arms (some dishwashers have upper and lower spray arms; others have a single arm), which hurl the cleaning agent against the dishes. When the wash cycle is completed, a rinse agent mixes with clean water and is circulated. The heating element can be called on to dry the dishes.

Troubleshooting

Problem   **Solution**

Dirty or spotted dishes after washing

Check manufacturer's loading recommendations • Experiment with detergent and rinse agent brands and amounts to determine if a variation is more effective • Test the selector switch **81** • Test the timer motor **82** • Inspect the detergent and rinse agent dispensers **85** • Check for and test the bimetal terminals on rinse agent dispenser **85** • Look for obstructions that block spray arm; check and clean the spray arm **87** • Test water temperature **88** • Test the heating element **88** • Check the water pressure; if low, avoid using house water supply while running dishwasher **89** • Check for clogged pump or damaged impeller; call for service •

Dishwasher doesn't fill with water

Check whether water flows from water supply • Adjust the door latch and test the door switch **83** • Inspect and clean the spray arm **87** • Inspect the float and test float switch **89** • Clean faucet coupler filter (portable models) • Test water inlet valve solenoid **90** •

Dishwasher drains during fill

Test the drain valve gate arm mechanism **91** • Test the drain valve solenoid **91** •

Dishwasher gets stuck in a cycle

Test the timer motor **82** •

Water doesn't shut off

Test the timer motor **82** • Inspect the float and test float switch **89** • Test the water inlet valve solenoid **90** • Remove the water inlet valve; clean or replace the valve screen **90** •

Motor doesn't run

Test the timer motor **82** • Adjust the door latch and test the door switch **83** • Test the motor **93** • Check for a blown fuse or tripped circuit breaker •

Motor hums, but doesn't run

Check for a blown fuse or tripped circuit breaker • Test the timer motor **82** • Adjust the door latch and test the door switch **83** • Test the motor **93** • Impeller jammed (older models) •

Troubleshooting

Problem	Solution

Poor water drainage — Test the timer motor **82** • Inspect the drain hose **86** • Inspect and clean the spray arm filter **87** • Check for reversing motor and test **91** • Test the drain valve solenoid **91** • Pump impeller clogged or broken (older models); call for service **92** •

Dishwasher leaks around door — Use detergent recommended for dishwashers only; do not prewash dishes with liquid detergent • Adjust the door latch **83** • Service the door springs **84** • Check and replace the door gasket **84** •

Dishwasher leaks from bottom — Tub cracked; seal crack with silicone rubber sealant or epoxy glue • Inspect the drain hose **86** • Check the spray arm for damage **87** • Tighten water inlet connection **90** • Pump seal faulty; call a service technician •

Dishwasher doesn't turn off — Test the cycle extender switch **82** • Test the timer motor **82** •

Dishwasher is noisy — Check and follow manufacturer's loading recommendations • Check the spray arm for damage **87** • Check water pressure; if low, avoid using house water supply while running dishwasher **89** • Check water inlet valve screen **90** • Test the inlet valve solenoid if you hear knocking sound during fill **90** •

Door drops hard when opened — Replace door springs or cable **84** •

Door is difficult to close — Adjust or replace door latch **83** •

TOOLS

Screwdriver
Adjustable wrench
Multitester
Hose-clamp pliers

MATERIALS

Masking tape
Shallow pan
Ruler

SAFETY FIRST

Built-in dishwashers are hard-wired into your home's electrical system. Before servicing any electrical components, always disconnect power to the machine at the house's main service panel.

Before You Start

Built-in dishwashers are installed under a kitchen counter with permanent plumbing and wiring connections. Portable models have a plastic coupler that connects to a sink faucet and drain, and a power cord that plugs into a 120-volt outlet.

HELPING DISHWASHERS DO THEIR BEST

The most common complaint about dishwashers is that they fail to clean dishes completely. While the cause may be an internal problem, very often the fault lies elsewhere.

If your dishwasher doesn't clean thoroughly, first check the owner's manual for the manufacturer's recommendations on loading, detergents, and rinse agents. If the water in your area is unusually high in minerals, try using more detergent; if you have a water softener, use less.

Determine whether the dishwasher is getting enough water (*page 89*). Low household water pressure may be to blame and may require intervention by the local water utility, a plumber, or a well specialist.

Most other problems result from electrical component faults and are easy to diagnose and repair. The most complex dishwasher problems involve the pump and motor. For example, a faulty pump seal may cause a leak into the motor, causing a failure. Even so, it's not difficult to test the motor, remove it for service, or to replace it.

Before You StartTips:

⋯⋗ It is rarely necessary to pull a built-in dishwasher out from under a counter to make repairs. On most models, all key parts are accessible through the front panels.

Access to the Controls

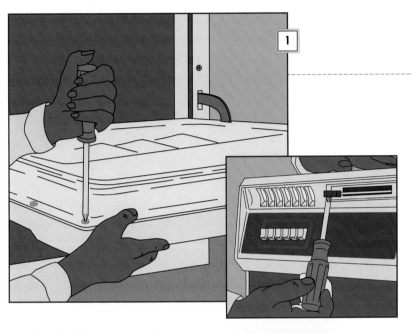

1. REMOVING RETAINING SCREWS

You will need to disassemble the door to service the selector switch, cycle extender switch, door switch, and timer.

● Unplug the dishwasher or turn off power at the service panel.

● Take out any retaining screws on the interior door *(left)* and on the front of the control panel *(inset)*.

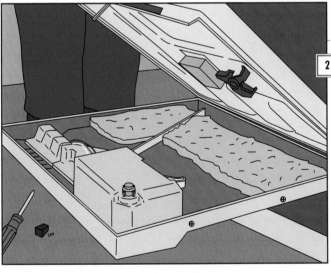

2. REMOVING THE DOOR PANEL

● Lift the interior door panel and set it aside to expose the rear of the control panel *(left)*.

3. ACCESSING THE CONTROL PANEL

● Take off any control-panel covers you find. To do so, remove the plastic clips that secure the covers to the door.

● Lift the control panel covers to expose control panel components *(left)*.

Access to Parts Beneath the Tub

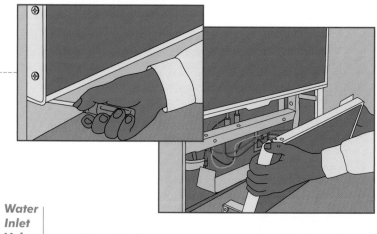

REMOVING THE LOWER PANEL

Many components can be reached only after the lower panel is removed.

● Turn off power to the dishwasher.

● Remove the retaining screws *(near right)* and free the panel by either pulling it down or lifting it off its hooks *(far right).*

● The illustration at bottom right shows the array of components found behind the lower panel of a typical dishwasher.

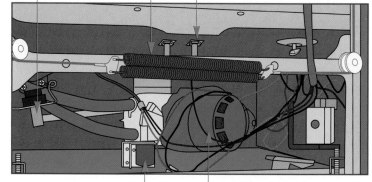

Water
Inlet
Valve   Door Spring   Heating Element Terminal

Drain Valve   Pump-and-Motor
Solenoid   Assembly

Ron's TRADE SECRETS

CLEANING AND PREVENTING RUST STAINS

If the walls inside your dishwasher are stained with rust, one cause can be a nick in the coating of the steel tub, or the coating on one of the racks may be chipped, exposing the steel wire to oxidation.

In either case, you've got to clean up the dishwasher and repair the problem at its source. I like to use oxalic acid solution—sold in home centers as deck cleaner or wood bleach—to scrub away the stains. That helps me find the nicks and scratches.

After taking the rusty spots down to bare metal—it's sometimes necessary to use emery cloth or another fine abrasive—I patch the surface with the type of plastic coating solution sold in hardware stores to protect and insulate tool handles.

Connections

Upper Spray Arm Hose

Drain Hose

Water Inlet Valve

DISCONNECTING THE POWER AND WATER

• Turn off the water and power at their respective sources.

• Find the water inlet valve and place a bowl or bucket beneath it.

• With a wrench or pliers, unscrew the compression fittings on the water supply line from the water inlet valve *(left)*.

• Disconnect the drain hose from the sink drain.

Selector Switch

TESTING FOR CONTINUITY

• Turn off the power to the dishwasher and expose the control panel *(page 79)*.

• Set a multitester to RX1 and disconnect the wires from one pair of terminals. Touch a probe to each switch terminal of the pair *(left)*. The meter should show continuity with the related switch button pressed *(page 18)*.

• Test each pair of terminals on the switch, removing and replacing wires as you go. Replace the entire unit if any pair fails, swapping wires to the new switch one by one to ensure correct connections.

Cycle Extender Switch

1. TESTING THE SWITCH

• Turn off power to the dishwasher and expose the control panel *(page 79)*.

• Label and disconnect the wires from their positions on the switch, which is located adjacent to the timer motor.

• Set a multitester to RX1; test terminals H2 and L2 *(right)* for continuity *(page 18)*.

• Set a multitester to RX1000; test terminals H2 and H1 for resistance of about 5400 ohms. Replace the switch if it fails either test *(next step)*.

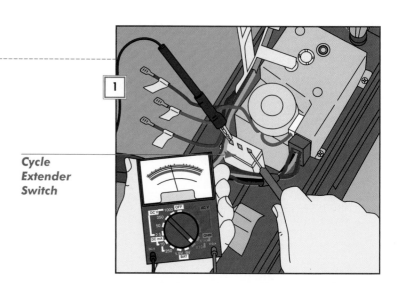

Cycle
Extender
Switch

2. INSTALLING A NEW SWITCH

• Unscrew or unclip the switch from the control panel *(right)*.

• Install a new switch; connect the wires one by one.

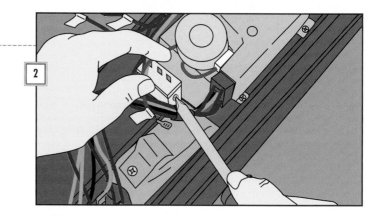

Timer Motor

1. TESTING THE WINDINGS

• Turn off power to the dishwasher and expose the control panel *(page 79)*, then disconnect the motor wires from the cycle extender switch.

• Set a multitester to RX1000, and probe the motor wire terminals *(right)*.

• Replace the timer if the resistance is not about 3,300 ohms *(next step)*.

Timer
Motor

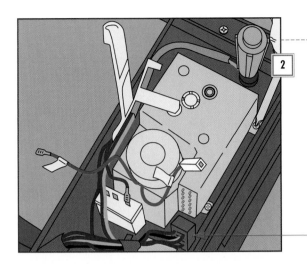

Timer Plug

2. REPLACING THE TIMER

• Remove the timer plug and inspect it. If you notice bent or protruding contacts, push them back into place.

• Label and disconnect, all of the motor wires. Pull the knob off the front of the timer, then unscrew it from the back of the control panel *(left)*.

• Install a new timer. Reconnect the motor wires, timer plug, and knob, then reattach the control panel.

Testing the Door Switch

1. TESTING THE DOOR LATCH

• To adjust the door latch, loosen the retaining bolts with a nut driver and slide the latch in and out. If the door latch closes securely and the machine will not run, test the switch *(next step)*.

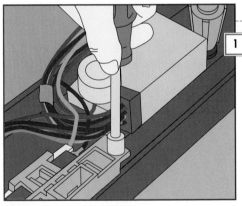

2. TESTING AND REPLACING THE DOOR SWITCH

• Turn off power to the dishwasher.

• Set a multitester to RX1 and test the switch for continuity *(left)*. The meter should indicate continuity with the switch button pushed in, an open circuit when it's out. Replace the switch if your test results are different.

• To replace the switch, remove the door switch retaining bolts *(previous step)* and remove the switch assembly.

• Install a new door switch and reconnect the switch wires.

Door Springs and Cables

INSPECTING THE SPRING, CABLE, AND PULLEY ASSEMBLY

If the door falls to a horizontal position when opened, examine the spring, pulley, and cable assembly.

● Turn off power to the machine and remove the lower panel *(page 80)*.

● Inspect the cables and springs; always replace damaged springs and cables in pairs to ensure proper tension.

● Grasp a damaged cable (or spring) with your hand and remove it from its hook.

● Close and lock the door. Reroute a new cable around the pulley, secure it to the door assembly, and reinstall the lower panel.

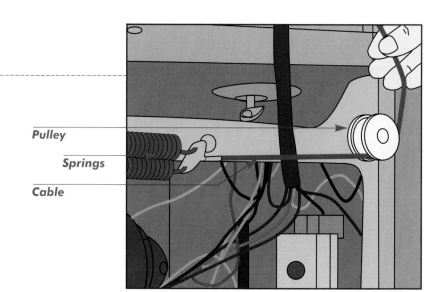

Pulley

Springs

Cable

Door Gaskets

REPLACING THE DOOR GASKET

● Open the dishwasher door and remove the dish racks.

● Inspect the gasket. If it is supple and un-cracked, adjust the door latch *(page 83)*. Otherwise, replace the gasket.

● Pry out the old gasket with a screwdriver and remove it *(right)*.

● Lubricate the new gasket with water only (don't use soapy water) and slide it into place.

● With your thumb or the handle of a screwdriver, press the center of the gasket into the top center of the door. Continue around the door, pressing it into place a few inches at a time in each direction.

Detergent and Rinse Agent Dispensers

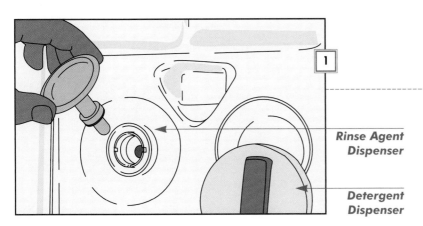

Rinse Agent
Dispenser

Detergent
Dispenser

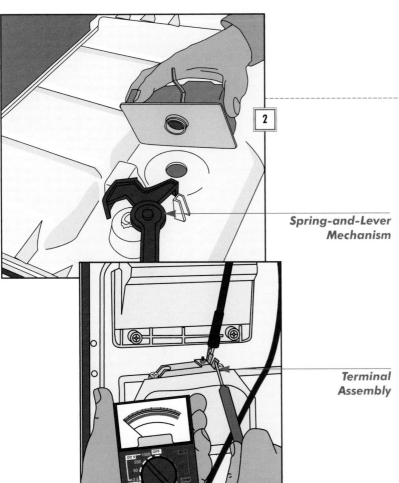

Spring-and-Lever
Mechanism

Terminal
Assembly

1. INSPECTING THE RINSE AGENT AND DETERGENT DISPENSERS

Caked-on detergent or rinse-agent residue may impede the release mechanisms. O-rings may fray or lose flexibility with age.

• Turn off power to the dishwasher. Check the detergent cup inside the door for caked-on residue *(left)*; clean it if necessary.

• Check the O-rings on the covers; replace them if damaged.

2. INSPECTING THE DISPENSER MECHANISMS

Most dishwashers have detergent and rinse-agent dispensers that are operated by a spring-and-lever mechanism. However, on some models, the rinse-agent dispenser is operated electrically. In either case:

• Begin by removing the interior door panel *(page 79)*.

• For a spring-and-lever mechanism *(left)*, release the plastic tabs and remove the rinse agent dispenser. Check the spring-and-lever mechanism for stuck or broken parts. Replace any part that is damaged.

• For an electrically-operated dispenser *(inset)*, set a multitester to RX1 and test for continuity. If the circuit is open, as shown here, unscrew the terminal assembly and replace it.

Drain Hose

1. CHECKING THE HOSE BENEATH THE WASHER

• Turn off power to the dishwasher and remove the lower panel *(page 80)*.

• Check that the clamp is secure and straighten any kinks by hand *(right)*. Replace the hose if any kinks remain.

• If there are no problems under the dishwasher, proceed to the next step.

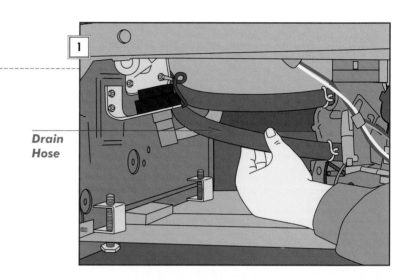

Drain
Hose

2. REPLACING THE HOSE

• With hose-clamp pliers, squeeze the spring clamp to disconnect the drain hose from the pump *(right)*. Have a shallow pan handy to catch any dripping water.

• Disconnect the hose at the kitchen sink drain or faucet coupler.

• Position the new hose, then reconnect it at both ends with new clamps.

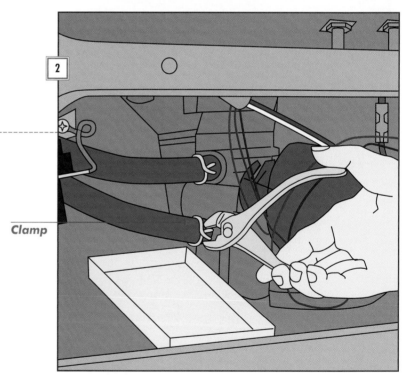

Clamp

Spray Arms

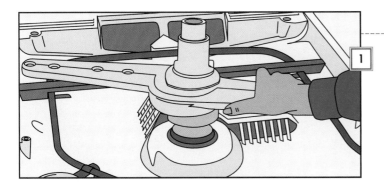

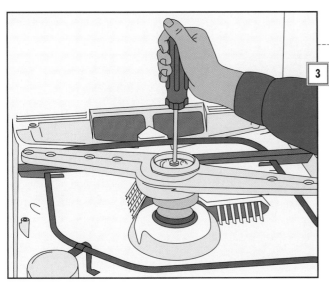

Spray Tower

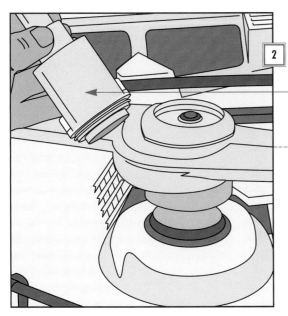

1. CHECKING ARM MOVEMENT

• Turn off power to the dishwasher and slide out the lower dish rack.

• With your hand, rotate the spray arm to see if it moves freely *(left)*. The ends of the arm should also move up and down slightly. Repeat this test on the upper spray arm, if present.

• Replace the arm if it doesn't spin freely or if it is bent or otherwise damaged. Most upper spray arms simply bolt or screw to the top of the tub. To remove the lower spray arm, proceed to Step 2.

2. DETACHING THE SPRAY TOWER

• Pull upward on the telescoping parts on the spray tower to see if they move freely.

• Unscrew the spray tower by hand and remove it *(left)*.

• Check for clogs and clean as necessary.

3. REMOVING THE SPRAY ARM

• Unscrew the bolt (if any) in the center of the spray arm *(left)*.

• Lift out the spray arm and its washers, keeping them in order.

• Clean the holes in the spray arm or replace it if damaged.

• Reuse the old washers if the new spray arm does not come with new ones.

• Remount the spray arm and spray tower.

Water Temperature

MEASURING WATER TEMPERATURE

• Turn on the dishwasher, then interrupt it during the first cycle by opening the door. Wait for the steam to clear before proceeding.

• Place a meat or candy thermometer in the water at the bottom of the dishwasher *(right)*. Look for a reading of at least 140°F.

• If the reading is too low, check if your household water heater is on a timer that limits when it will heat household water.

• If the water heater is set properly, check the heating element *(below)*.

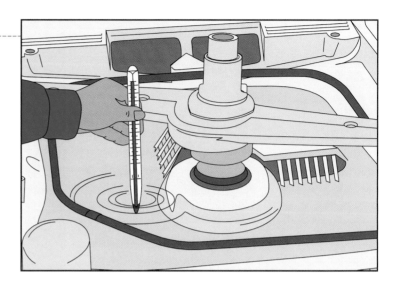

Heating Element

1. TESTING THE HEATING ELEMENT

• Turn off power to the dishwasher and remove the lower panel *(page 79)*.

• Set a multitester to RX1, disconnect one of the wires from the element, and touch one probe to each terminal *(right)*. Look for a reading different from 0 or infinity.

• Place one probe on a terminal and the other on the element's metal sheath, and check for ground *(page 18)*. Replace the element if it fails either test.

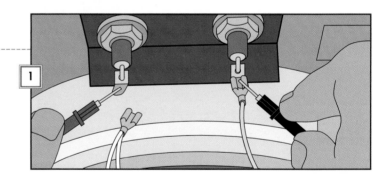

2. INSTALLING A NEW ELEMENT

• Slide off the rubber terminal cover (if any) and remove the lock nuts securing the element under the tub *(right)*.

• From inside the tub, lift out the element *(inset)* and replace it. Tighten the lock nuts and reconnect the wires to the terminals.

Lock Nut

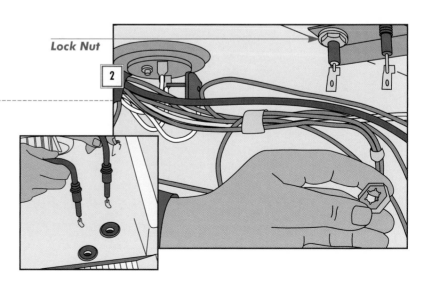

Water Pressure

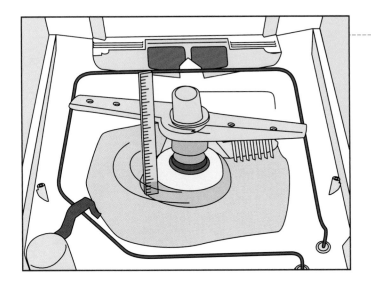

MEASURING THE WATER LEVEL

Water flows to a dishwasher in a timed cycle. If water pressure is too low, the tub won't fill enough for thorough cleaning.

• Turn on the dishwasher; stop it midway through the first cycle. Open the door and let the steam clear.

• Stand a ruler on the sump. If the water level does not reach the 3/8-inch mark on the ruler, avoid using water elsewhere in the house when running the dishwasher.

Float Switch

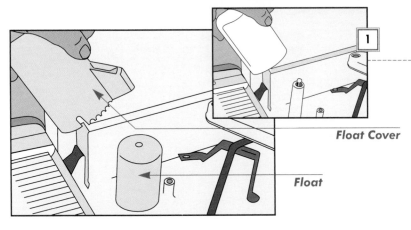

Float Cover

Float

1. INSPECTING THE FLOAT

• Turn off power to the dishwasher, open the door, and remove the lower rack.

• Remove the float cover *(left)*. Jiggle the float up and down to see if it moves freely.

• Pull off the float *(inset)*. Look for obstructions. If you find none, check the float switch *(next step)*.

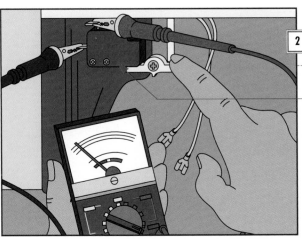

Float Switch

2. TESTING THE FLOAT SWITCH

• Remove the lower panel *(page 80)*. Find the switch directly below the float, then detach the wires from its terminals.

• Set a multitester to RX1 and clip a probe to each terminal.

• Pull down on the lever to put the switch in the ON position. A faulty switch will show no continuity *(left)*; replace it.

Water Inlet Valve

1. CHECKING FOR CONTINUITY

• Remove the lower panel *(page 80)*. Make sure the incoming water line and the hose that connects the inlet valve to the tub are securely fastened.

• Remove the wires from the inlet valve terminals. Set a multitester to RX1.

• Touch a probe to each terminal *(right);* a good valve will show some resistance.

• Remove and repair the valve if the meter reads infinity *(next step).*

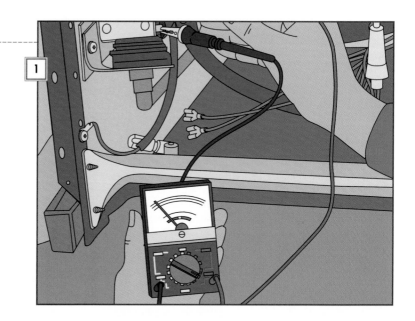

2. DISCONNECTING THE HOSE

• Loosen the hose clamp that connects the inlet hose to the valve *(right).*

• Disconnect the incoming water line.

• Remove any screws that secure the valve bracket to the side of the dishwasher. Replace the valve if the filter screen is clogged or if it shows any signs of damage.

• Reinstall the valve, water line, and hose, then check all connections.

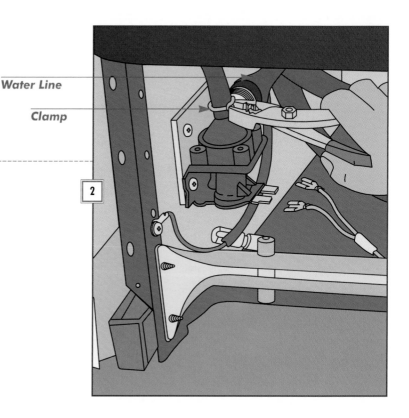

Water Line

Clamp

Drain Valve

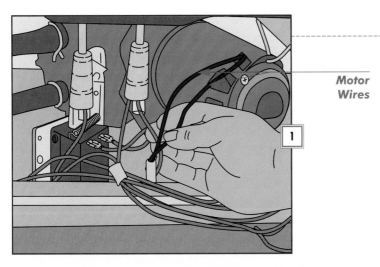

Motor Wires

1. LOCATING THE DRAIN VALVE

Only dishwashers with nonreversible motors have a drain valve.

• Turn off power to the dishwasher and remove the lower panel *(page 80)*.

• Count the number of wires attached to the motor. A motor with two or three wires is nonreversible *(left)*; one with four wires is reversible and requires no drain valve.

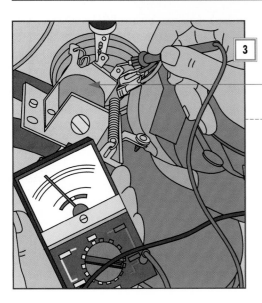

Gate Arm

Drain Valve

2. CHECKING THE GATE ARM MECHANISM

• Move the gate arm mechanism by hand *(left)*; it should move freely up and down.

• Inspect both springs; replace them if they are missing or broken.

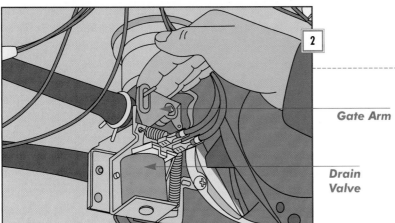

Drain Valve Solenoid

3. TESTING THE DRAIN VALVE SOLENOID

• Disconnect the wires from the drain valve solenoid. Set a multitester to RX1.

• Clip a probe to each terminal on the solenoid *(left)* and check for about 40 ohms of resistance *(page 18)*.

• If the tester reads infinity, indicating an open circuit, replace the solenoid *(next page)*.

4. REPLACING THE SOLENOID

• Note carefully how the mounting screws, springs, and wires attach to the solenoid.

• Detach the solenoid springs, then remove the mounting screws *(right)*.

• Screw a new solenoid in place, then reattach the springs and the wires.

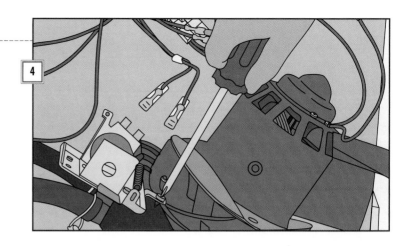

Pump and Motor Assembly

1. INSPECTING THE MOTOR

• Turn off the power and remove the lower panel *(page 79)*.

• Try to turn the motor fan blades by hand *(right)*. Use a screwdriver if you can't reach in with a finger. If they don't move freely, look for obstructions or call for service.

Ron's TRADE SECRETS

FIGHTING FOOD-PARTICLE FILM
If your dishes have a light film of food particles on them after a full session in your dishwasher, the problem may be a clogged sink drain or malfunctioning garbage disposer. Waste water from the dishwasher usually empties into the same drainpipe as the kitchen sink and disposer. If the sink drain or disposer is clogged, the dishwasher's dirty water can't drain, causing food particles to redeposit on the dishes.

Try clearing the sink drain with a plunger or auger. Then check that the sink drains freely. If the sink drain and disposer are clear, check the dishwasher's drain hose. If it's kinked, pinched, or clogged, that might also impede drainage and cause film to form. Remove any debris lodged in the hose, straighten out kinks, or replace the drain hose, as needed.

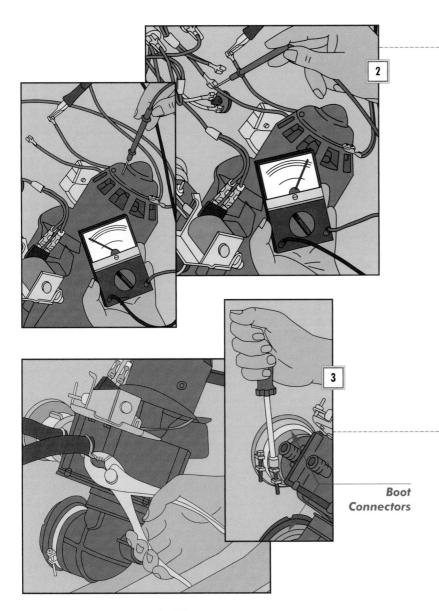

2. TESTING THE MOTOR

• Snap off any protective brackets, then disconnect the motor wires from their terminals.

• Set a multitester to RX1 and attach a probe to each motor wire terminal *(near left)*. The motor should show about 2 ohms of resistance *(page 18)*.

• If the test indicates full continuity (the needle swings to 0), test for a ground. To do so, place one probe of the multitester on the bare metal housing of the motor, and the other probe on each terminal in turn *(far left)*. The needle should not move.

• If the motor fails either test, proceed to the next step.

3. DISCONNECTING THE PUMP-AND-MOTOR ASSEMBLY

• With hose-clamp pliers, disconnect the hoses from the pump *(far left)*. Mark the hose positions for reconnection.

• Detach the drain-valve solenoid wires.

• Unscrew the clamp closest to the pump on each boot connector *(near left)*.

Boot
Connectors

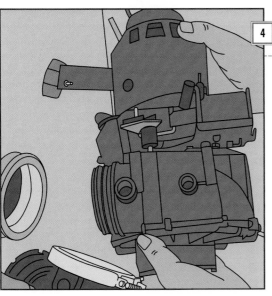

4. REPLACING THE PUMP/MOTOR

• Remove the ground wire from the assembly, then pull the assembly free of its hanger *(left)*.

• Replace the assembly or take it to a service center for repair.

• To reinstall the motor assembly, remount it to its brackets, reattach the hoses, and reconnect the wires.

FIX IT: Garbage Disposers

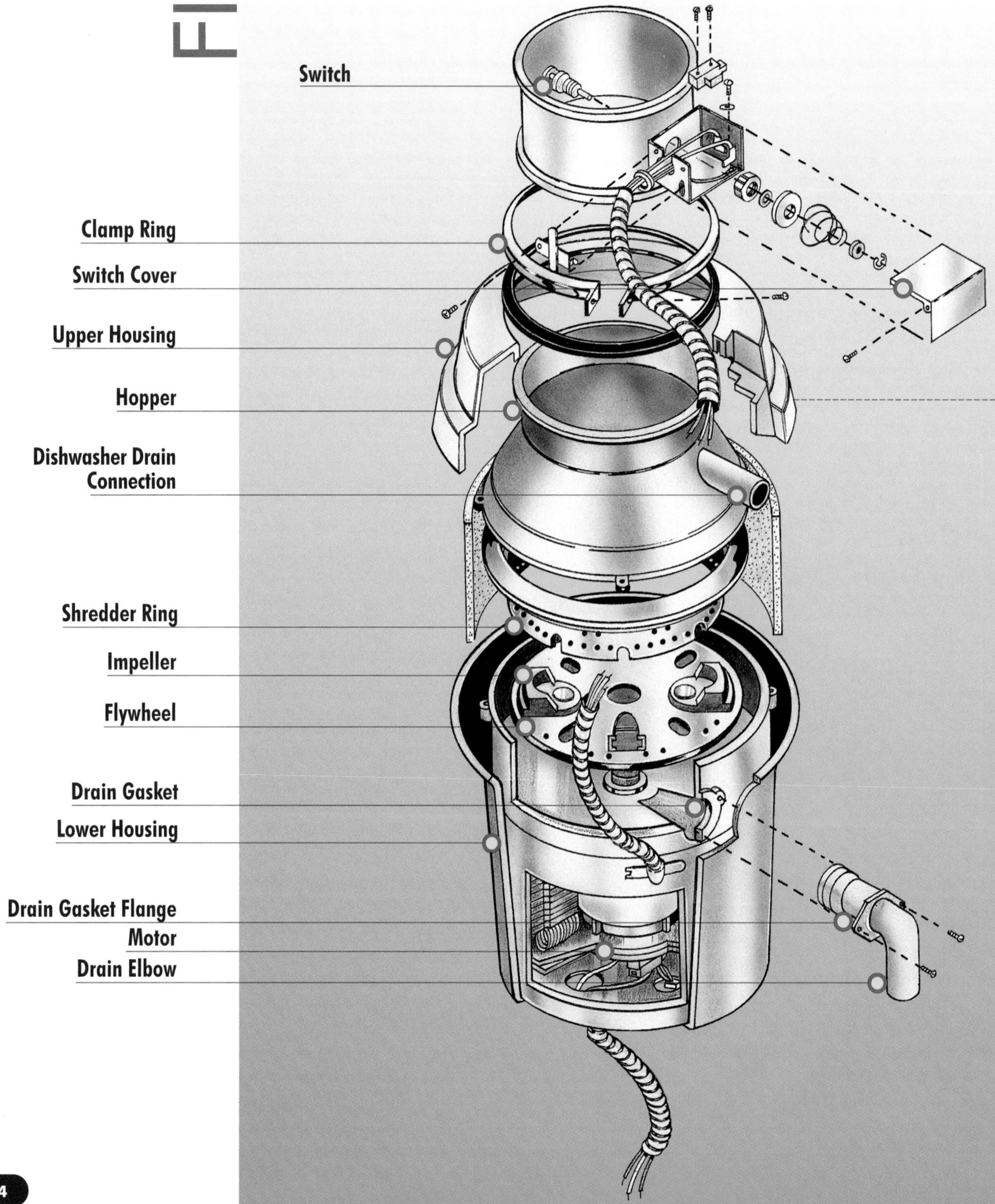

Switch

Clamp Ring

Switch Cover

Upper Housing

Hopper

Dishwasher Drain
Connection

Shredder Ring

Impeller

Flywheel

Drain Gasket

Lower Housing

Drain Gasket Flange

Motor

Drain Elbow

Chapter 5

Contents

How They Work

Garbage is fed into the disposer through the sink drain and collects in a hopper with a flywheel at the bottom. In some units, waste also comes through a hose from the dishwasher drain. When the disposer is activated, a motor spins the flywheel. Waste speeds outward to the shredder ring, where it is ground into small pieces with the help of impellers. Running water flushes the ground-up waste through openings in the flywheel and down the drain.

Garbage disposers differ mainly in the way they are switched on and off. The illustration at left shows the parts of a typical batch-feed garbage disposer. Batch-feed units are activated by a switch that is tripped by the stopper. Continuous-feed disposers are controlled by a wall switch.

Troubleshooting

Problem	Solution
• **Disposer won't run at all**	Check for tripped circuit breaker or blown fuse • Check for jam or tripped overload protector **98** • Test wall switch (continuous-feed disposers) **101** • Test integral switch (batch-feed disposers) **101** • Motor faulty; call for service •
• **Disposer drains poorly**	Open cold water faucet fully when operating the disposer • Dismount the disposer **99** • Check and clear the trap **99** • Use a plumber's auger to clear the drain or call a plumber (do not use chemical drain openers) •
• **Disposer will not stop**	Test wall switch (continuous-feed models) **101** • Test switch (batch-feed models) **101** •
• **Disposer grinds too slowly**	Open cold water faucet fully when operating the disposer • Disconnect the disposer and remove any waste not recommended for disposal by the manufacturer •
• **Disposer starts, but stops when stopper is released (batch-feed models)**	Test the switch **101** • Inspect the cam; replace if necessary **101** •
• **Disposer leaks**	Dismount the disposer **99** • Remove the drainpipe and replace the gasket **99** • Tighten screws or bolts on mounting ring **100** • Tighten the drain gasket screws **100** •
• **Disposer vibrates or is unusually noisy**	Check for objects in hopper • Tighten loose mounting screws or bolts **100** • Motor is faulty; call for service •

Before You Start

Garbage disposers make short work of shredding tough food wastes such as melon rinds and chicken bones. Used properly, these rugged appliances typically serve a decade or more without complaint.

HINTS FOR TROUBLE-FREE SERVICE

To avoid problems with a garbage disposer, be careful about the amount and types of waste that you place in it for processing. Generally speaking, however, it's best to put waste in loosely; packing can jam the machine. Always run cold water when operating the disposer to assist shredding and congeal grease. Never subject the disposer to clam shells and similar, hard-to-grind waste or to non-food waste; all can damage the unit. Cut or break up large bones, corncobs, and fibrous vegetable wastes before putting them into the disposer.

When the disposer is not in use, keep the drain cover in place to prevent objects from falling into the unit. Don't leave waste in the disposer for more than 24 hours. If the disposer emits odors, try grinding some orange or lemon rind. Never use a chemical drain opener to clear clogs—it can damage the plastic and rubber parts.

Before You StartTips:

⋯⟩ If the disposer doesn't run, first check the main service panel for a blown fuse or tripped circuit breaker.

⋯⟩ Garbage disposers weigh as much as 15 pounds. When dismounting the unit, be prepared to support the weight.

TOOLS

Screwdrivers
Hex keys
Wrenches
Nut drivers
Hose-clamp pliers
Multitester

MATERIALS

Tongs
Broom handle
Bucket

SAFETY FIRST

Never put your hand down a disposer, even when it's off; you might activate the switch. Use tongs to retrieve objects that have fallen into the hopper.

Before attempting any repair, first turn off power at the main service panel.

Clearing Jams

RESETTING THE OVERLOAD PROTECTOR

• Disconnect power to the disposer.

• If the jam was caused by a utensil or other object, use tongs to remove it and wait 15 minutes for the motor to cool.

• Reconnect the power, push the overload protector reset button on the bottom of the disposer *(right),* and try operating the disposer again.

UNBLOCKING THE FLYWHEEL

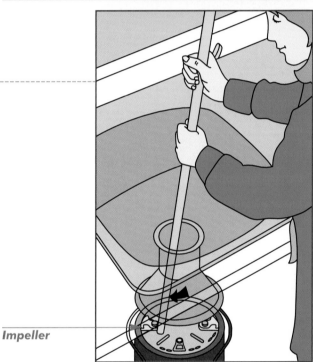

• Disconnect power to the disposer.

• Insert the end of a broom handle into the disposer and wedge it against one of the impellers on the flywheel. Force the wheel back and forth until it moves freely *(right).*

• With tongs, remove any objects that may become dislodged.

• Reconnect the power, push the overload protector reset button, and try operating the disposer again.

Impeller

ROTATING THE MOTOR SHAFT

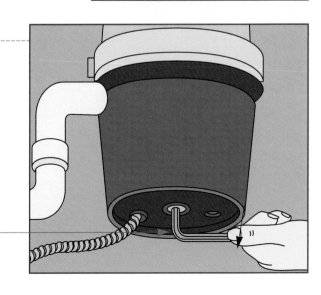

• If your disposer has a hexagonal fitting at the bottom, turn off power to the disposer and insert a hex key. Turn it back and forth until the flywheel turns freely *(right).*

• With tongs, remove any objects jamming the disposer and wait 15 minutes for the motor to cool.

• Reconnect the power, push the overload protector reset button, and try operating the disposer again.

Hex Key

Connections

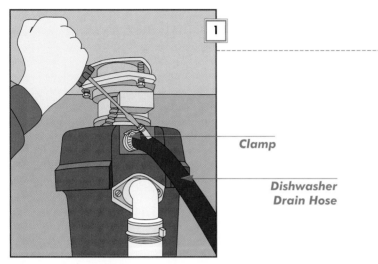

Clamp

Dishwasher Drain Hose

1. DETACHING THE DISHWASHER DRAIN HOSE

Not all units have a dishwasher drain hose.

- Shut off power to the disposer at the main service panel.

- Place a bucket or pan under the unit to catch any dripping water.

- With a screwdriver, loosen the clamp on the dishwasher drain hose *(left)*. Some units require hose-clamp pliers instead of a screwdriver.

- Pull off the hose.

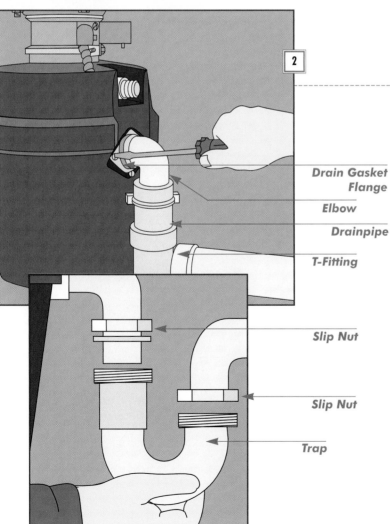

Drain Gasket Flange

Elbow

Drainpipe

T-Fitting

Slip Nut

Slip Nut

Trap

2. DISCONNECTING THE DRAINPIPE

A disposer may be connected to the drain line by means of a T-fitting or directly to the trap.

- Place a bucket under the unit to catch any dripping water.

- If the drain elbow leads to a T-fitting, unscrew the drain gasket flange *(left)*, loosen the fitting connecting the elbow to the drainpipe, and pull out the elbow.

- If the elbow from the disposer joins directly to the trap, loosen the slip nuts with pliers or a pipe wrench and pull the trap downward *(inset)*.

- Inspect the trap for any clogs or build-up and clean it with an auger and a flexible brush, if necessary.

3. DISMOUNTING THE DISPOSER

• If your disposer is fastened with a twist-lock mechanism, rotate it—usually to the left—until it slips free. Otherwise, loosen the mounting screws or bolts at the top of the disposer.

• Twist the unit free and lower it from the support ring *(right)*.

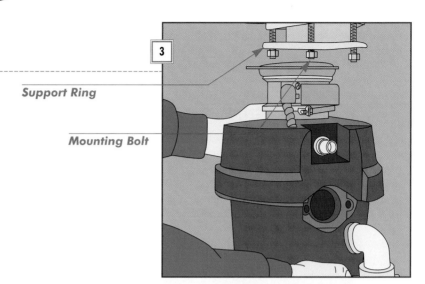

Support Ring

Mounting Bolt

4. DISCONNECTING THE POWER SUPPLY

If your disposer is not hard-wired, simply unplug it from the wall outlet. If your disposer is hard-wired, as most are, follow these steps.

• Unscrew the cover plate on the bottom of the disposer.

• Disconnect the green ground wire from its screw terminal.

• Disconnect the black and white wires by unscrewing the two wire caps *(right)*.

• Loosen the setscrew on the Bx connector and pull the cable free from the disposer.

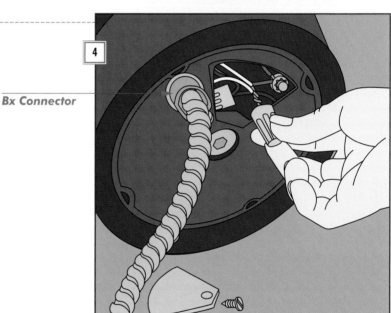

Bx Connector

Leaks

CHECKING THE DRAINPIPE AND MOUNTING GASKETS

• To repair leaks, first tighten the drain gasket flange *(page 99)*. If leaks persist, remove the drainpipe and replace the gasket.

• Tighten the bolts or screws on the support ring *(right)*.

• If leaking continues, disconnect the disposer *(page 99),* remove the support ring and mounting flange, and replace the sink flange gasket.

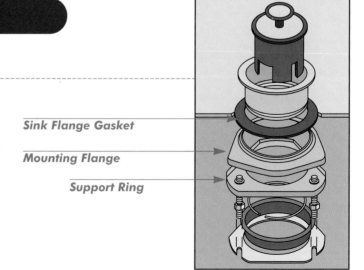

Sink Flange Gasket

Mounting Flange

Support Ring

Switches

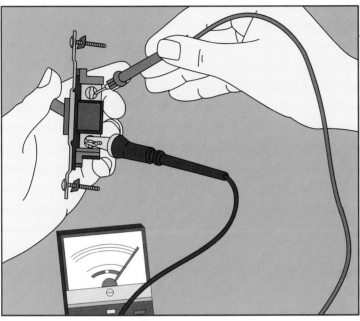

TESTING A WALL SWITCH

Continuous-feed disposers are controlled by a wall switch.

• Turn off power at the main service panel.

• Remove the cover plate, unscrew the switch from the box, and pull the switch from the wall.

• Disconnect the wires from the terminals and set the switch to ON.

• Set a multitester to RX1. Touch the probes to the switch terminals *(left)*, and test for continuity *(page 18)*.

• If the switch fails the test, replace it.

TESTING AN INTEGRAL SWITCH

Batch-feed disposers are controlled by an integral switch.

• Turn off power at the main service panel and dismount the disposer *(page 99)*.

• Unscrew the switch cover.

• Disconnect the terminal leads and set the switch to the ON position by inserting the stopper and turning it until it will not go any farther.

• Set a multitester to RX1. Touch the probes to the switch terminals *(top left)*, and test for continuity *(page 18)*.

• If the switch fails to show continuity, unscrew it from the disposer and replace it with a new one.

• If the switch is not faulty, check the cam for wear or damage *(bottom left)*. If its condition prevents it from activating the switch button, replace the stopper.

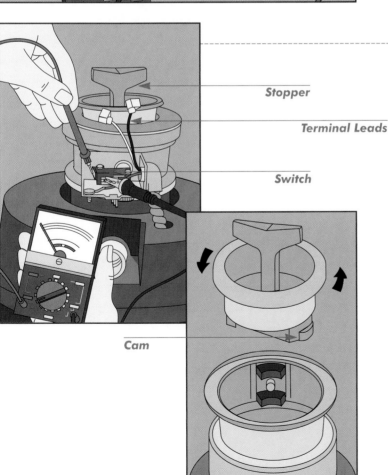

Stopper

Terminal Leads

Switch

Cam

FIX IT: Clothes Washers

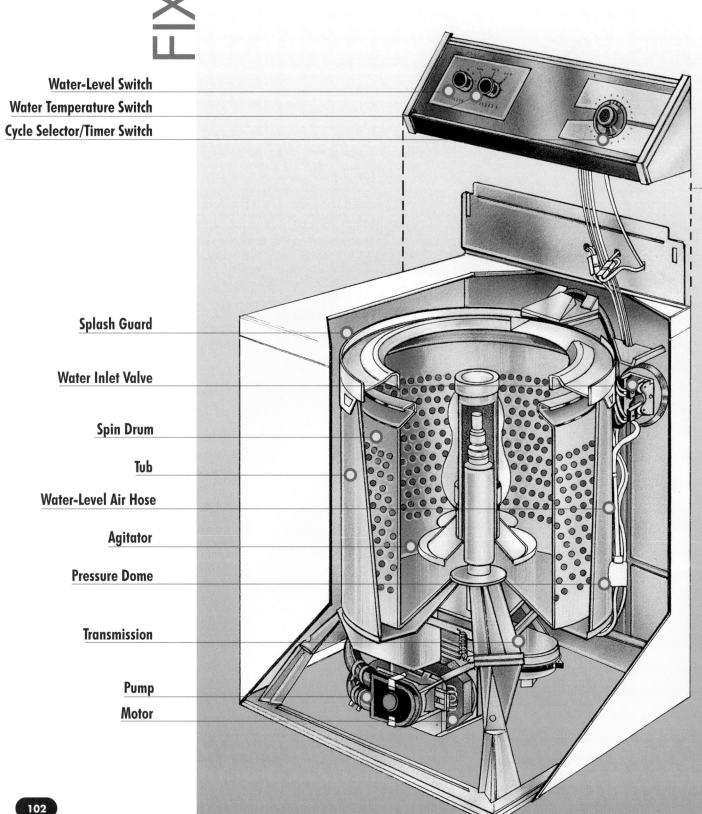

Water-Level Switch

Water Temperature Switch

Cycle Selector/Timer Switch

Splash Guard

Water Inlet Valve

Spin Drum

Tub

Water-Level Air Hose

Agitator

Pressure Dome

Transmission

Pump

Motor

Chapter 6

Contents

How They Work

In all washing machines, knobs on a control panel send wash-cycle instructions to a timer and electrically operated inlet valves. A wash cycle begins by filling the washtub with hot, cold, or warm water that also permeates a spin drum, which holds the clothes for washing. On cue from a timer, an agitator at the center of the spin drum churns clothes to clean or rinse them. Then the drum spins to extract water from the clothes and a pump channels the wastewater down the drain.

For all the similarities among models in use today, clothes washers differ in how the motor is connected to the agitator and spin drum. In the direct-drive type shown here, these two components are connected directly to the motor shaft by means of bearings and gears. In belt-drive models (page 109), power from the motor is transferred to the agitator and spin drum by a belt-and-pulley system.

Troubleshooting

Problem	Solution
• **Washer doesn't run at all (in some cases, motor may hum)**	Check that the washer is plugged in; check for a blown fuse or tripped circuit breaker • Test the water level switch assembly **110** • Test the timer motor **111** • Motor overheated; turn off washer and allow it to cool for one hour • Test the lid switch **113** • Inspect the pump for blockages **117** • Test the centrifugal switch **118** • Test the motor **119** •
• **Washer doesn't fill**	Check the water supply hoses for kinks • Test the water level switch assembly **110** • Test the water inlet valve **116** • Test the water temperature switch **112** • Test the timer motor **111, 112** •
• **Washer doesn't stop filling**	Test the water level switch assembly **110** • Unplug washer; if it stops filling, test the timer motor **111** • Test the water inlet valve **116** •
• **Washer doesn't agitate**	Inspect the agitator dogs **114** • Inspect the agitator and the agitator clutch **114** • Test the water level switch **110** • Examine the coupling **119** • Replace the transmission **120** • Service the drive belt system **121** •
• **Washer doesn't drain**	Straighten the drain hose; replace it if damaged • Test the timer motor **111** • Inspect the pump for blockages **117** •

Troubleshooting

Problem	Solution
• **Washer doesn't spin**	Test the lid switch **113** • Inspect the agitator and the agitator clutch **114** • Test the timer motor **111, 112** • Inspect the pump for blockage **117** • Examine the coupling (direct-drive washers) **119** • Replace the transmission **120** • Test the centrifugal switch **118** • Test the motor **119** • Service the drive belt system **121** •
• **Washer leaks**	Inspect all hoses; tighten hose clamps; replace damaged hoses • Clean the water supply hose filters **116** • Inspect the tub **116** • Inspect the pump for blockage **117** •
• **Washer is noisy or vibrates excessively**	Redistribute clothes to correct load imbalance • Adjust leveling feet • Inspect the pump for blockage **117** • Examine the coupling (direct-drive washers) **119** • Replace the transmission **120** • Adjust or replace the drive belt **121** •
• **Washer damages clothing**	Check the owner's manual for use recommendations • Inspect the agitator for cracks **114** • Touch up rust spots on spin drum **115** •
• **Washer leaves lint on clean clothing**	Wash lint producers like towels and other cottons separately from synthetics and permanent-press fabrics, which attract lint • Clean water supply hose filters **116** • Inspect the pump for blockage **117** •

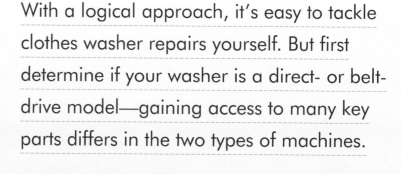

Before You Start

TOOLS

Screwdriver

Nut driver

Multitester

Socket wrench

Hose-clamp pliers

Spanner wrench

Hammer or rubber mallet

MATERIALS

Masking tape

Wooden block

Bucket

SAFETY FIRST

You may have a belt-drive washer that must be pulled away from the wall and tilted onto its back to expose the transmission, pump, or motor. The washer is extremely heavy. Do not attempt to tilt the machine yourself; enlist a helper.

With a logical approach, it's easy to tackle clothes washer repairs yourself. But first determine if your washer is a direct- or belt-drive model—gaining access to many key parts differs in the two types of machines.

How Machines Differ

While all clothes washers have similar electrical- and water-system parts, two wholly different drive systems are prevalent. The drive system—or the need for convenient access points to the specific type of motor and transmission assembly—largely determines the design of the housing.

Differences in housing designs sometimes create different paths for gaining access to switches, timers, valves, and the like. And by convention, the designs of certain switches and pumps differ in direct- and belt-drive models, requiring slightly different service procedures.

This chapter presents procedures for servicing both types of washers, making no distinction where access and parts have common schemes. But where there are significant differences, the section heading identifies the type of washer to which the instructions that follow apply. Look at the illustrations under each heading to determine whether the details closely resemble the configuration of your machine.

Before You StartTips:

⋯⋗ It's usually better to replace a faulty motor or transmission rather than to attempt repairs yourself.

Access to Direct-Drive Washers

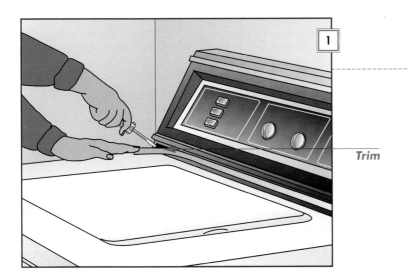

1

Trim

1. UNSCREWING THE CONTROL CONSOLE

- Disconnect power to the washer.

- Remove the retaining screws from the bottom corners of the console. The screws may be covered by an adhesive trim strip *(left)*. Some washers may have screws on the top or back of the console.

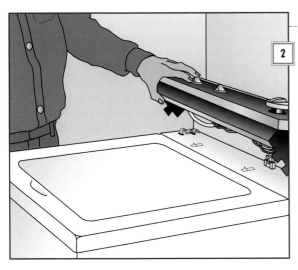

2

2. TIPPING BACK THE CONTROL CONSOLE

- Flip the console backwards to expose the timer, the timer motor, the water level switch, the water temperature switch, and the selector switch.

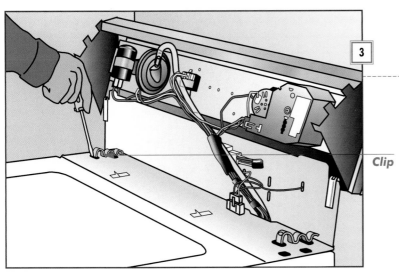

3

Clip

3. REMOVING THE CABINET CLIPS

- With a screwdriver, pry off the two retaining clips on either side of the washer. To do so, stick the screwdriver into the clip opening and rest it in a U-shaped trough *(left)*. Then, push the screwdriver handle rearward to dislodge the clip.

4. REMOVING THE LID SWITCH HARNESS

• Pull out the plug for the lid-switch wiring harness to disconnect the control console from the washer housing. If necessary, use a screwdriver to help loosen the clasp that secures the plug.

 Never disengage a plug by pulling on it.

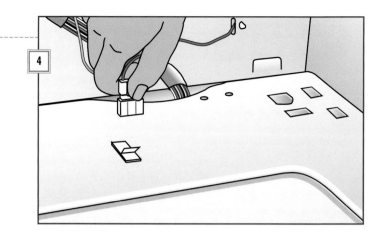

5. PULLING THE HOUSING FORWARD

• Tilt the housing forward at a 45-degree angle to disengage the tabs securing the washer to the frame *(right)*.

• Maintaining a 45-degree angle, pull the housing straight out and set it aside.

• To reinstall the housing when repairs are complete, hook the front edge under the front bottom rail, then lower the housing onto the side rails.

• Snap the tabs into cutouts in the housing.

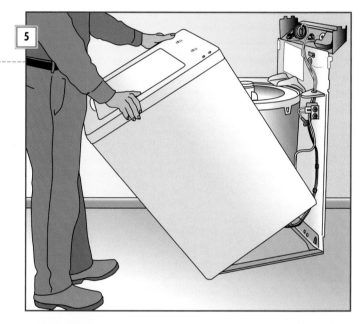

 Ron's **TRADE SECRETS**

THE RIGHT MOVES FOR DIRECT-DRIVE WASHERS
The housing provides most of the structural rigidity in direct-drive machines. After removing the housing, I'm always careful not to damage the rails it rests on at the front and sides. It's easy to step on them accidentally.

If you attempt to shift or lift the washer by grasping the rails, as shown in the photo at right, you're bound to do irreparable damage—you'll never get the housing to seat properly on the frame again.

If you need to move the washer after its housing has been removed, grasp only parts of the super-structure—the heavier metal struts that run from the center down to the corners.

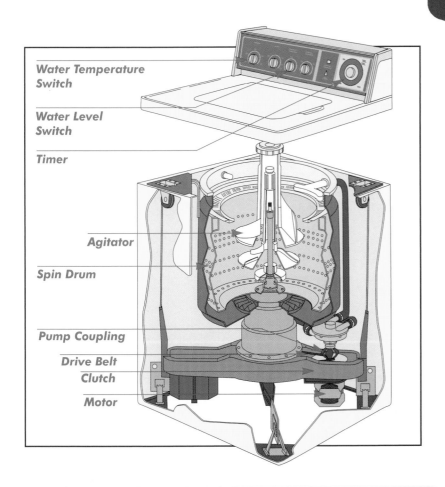

Water Temperature Switch

Water Level Switch

Timer

Agitator

Spin Drum

Pump Coupling

Drive Belt

Clutch

Motor

ANATOMY OF A BELT-DRIVE WASHER

The illustration at left shows the layout of the most common type of belt-drive washer. Most of its parts are similar to those on a direct-drive washer *(page 102)*, and can be found in similar locations. However, the motor-transmission-pump assembly is altogether different. In a belt-drive washer, the motor connects to the pump with a flexible coupling; and to the transmission by means of a rubber belt.

Access to Belt-Drive Washers

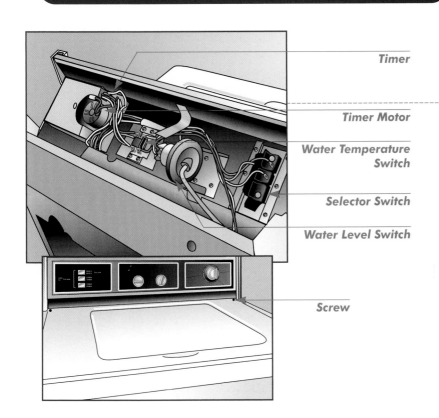

Timer

Timer Motor

Water Temperature Switch

Selector Switch

Water Level Switch

Screw

ACCESS THROUGH THE CONTROL PANEL

• Remove the screws from the corners of the control console *(inset)*.

• Tilt the console forward (the console at left is shown from the rear) to expose the timer and its motor, as well as the water-temperature, water-level, and cycle-selector switches.

ACCESS THROUGH THE TOP

• Tape the blade of a putty knife. Slip it between the top and the cabinet, then push inward to release the two spring clips securing the top of the washer. On some belt-drive washers, the spring clips are located at the front *(right)*. Others have the spring clips on the sides.

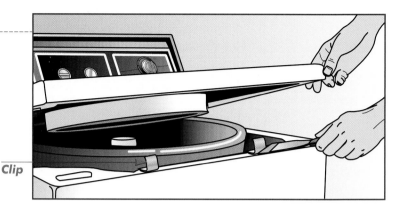

Clip

ACCESS THROUGH THE BACK PANEL

On belt-drive washers, you can usually expose the motor, pump, water inlet valve, and drive belt by removing the back panel *(right)*.

• Unplug the washer and disconnect the hoses from the water supply. Pull the appliance away from the wall.

• With a nut driver, take out the screws that secure the back panel *(inset)*.

• Remove the panel and set it aside.

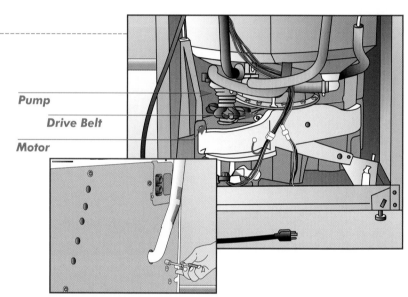

Pump

Drive Belt

Motor

Water Level Switch Assembly

1. INSPECTING THE AIR HOSE AND AIR PRESSURE DOME

• Unplug the washer and open the control console *(page 107 or 109)*.

• Remove the top or housing, as needed. Inspect the water level air hose for wear and straighten any kinks. Look for signs of water leakage on or around the hose and air pressure dome *(right)*.

• Check that all air hose connections are airtight.

• Replace the pressure dome if cracked or if the seal between it and the tub is broken.

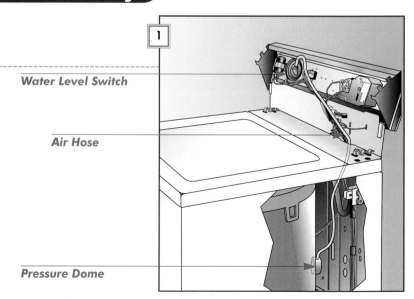

Water Level Switch

Air Hose

Pressure Dome

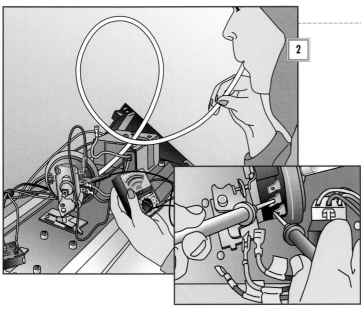

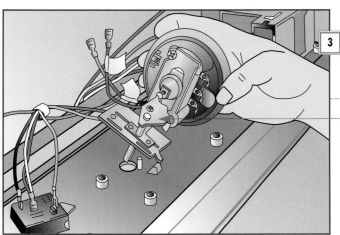

Bracket

2. Testing a water level switch

• Label and remove the three wires from the water level switch and push them aside. Set a multitester to RX1.

• Pull the air hose off the switch and attach a shorter tube of the same diameter. Gently blow into the tube *(left)*. Listen for a click as the switch trips to the FULL position.

• Continue blowing and test all three pairings of the switch terminals *(inset)*. Two pairs should show continuity, one should show resistance *(page 18)*; note results.

• Stop blowing and retest the switch. The terminal pair that showed resistance earlier should show continuity, and those that showed continuity should now show resistance. Replace the switch if the results differ *(next step)*.

3. Replacing a water level switch

• Pull off the water level knob, if any.

• Remove the air hose from the switch and unscrew its bracket *(left)*.

• Fasten the bracket of the new switch to the console and reconnect the air hose and wires. Push the knob onto the shaft.

Timers

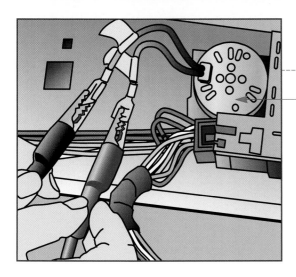

Timer Motor

Checking the motor (direct drive)

• Unplug the washer and open the control console *(page 107)*.

• Label and remove the wires leading to the timer motor.

• Set a multitester to RX100 and clip a probe to each terminal *(left)*. Look for a resistance reading of 2,000 to 3,000 ohms *(page 18)*. Replace the timer motor if the test result is different.

TESTING TIMER SWITCHES (BELT DRIVE)

• Unplug the washer and open the control console *(page 109)*.

• Release the plastic tab and lift off the timer switch cover. Check the wiring diagram and timer chart to identify the switch functions; label and disconnect the wires, omitted here for clarity.

• Turn the selector knob to the malfunctioning cycle.

• Set a multitester to RX1 and touch probes to the corresponding terminals *(right)*. If test readings for a switch differ from those listed in the accompanying table, the switch is faulty; call for service.

CONTROL KNOB POSITION	SWITCH					
	1. MOTOR	2. SPECIAL FUNCTION	3. MAIN POWER	3B. BYPASS	4. WASH	4B. SPIN
OFF	OPEN	(SEE	OPEN	OPEN	OPEN	OPEN
WASH	OPEN	TIMER	CLOSED	OPEN	CLOSED	OPEN
SPIN	OPEN	CHART)	CLOSED	CLOSED	OPEN	CLOSED

Water Temperature Switch

1. TESTING THE SWITCH

The water temperature switch may be either rotary or push-button. Both are tested the same way.

• Unplug the washer and remove the control console *(page 107 or 109)*.

• Check the wiring diagram on the washer for the markings used on the terminals that control the inoperative setting. Label and disconnect the wires from all terminals.

• Set a multitester to RX1 and set the switch to the inoperative setting. Touch probes to terminals *(right)*. The tester reading should indicate continuity *(page 18)*.

• If open, replace the switch *(next step)*.

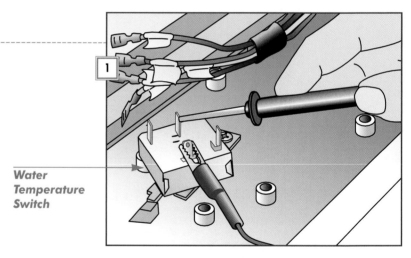

Water Temperature Switch

2. REMOVING THE SWITCH

• Unscrew the old switch from the console panel *(right)*.

• Install a new switch and connect the wires.

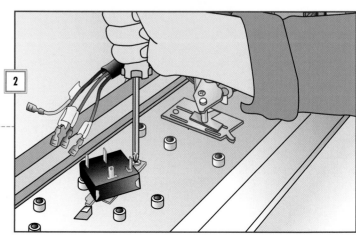

Lid Switch

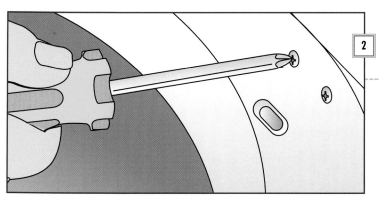

Wiring Harness

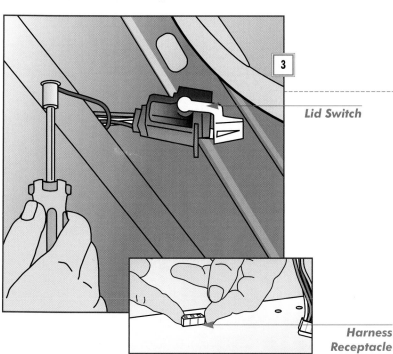

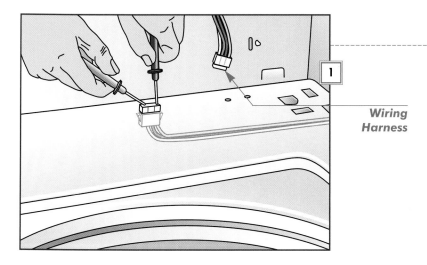

Lid Switch

Harness Receptacle

1. TESTING THE SWITCH

For safety, the lid switch turns off the washer when you open the lid.

● Unplug the washer and remove the the housing *(direct-drive, pages 107-108)* or the top *(belt-drive, page 110)*.

● Unplug the switch wiring harness. Set a multitester to RX1 and touch a probe to each switch terminal *(left)*. The meter should point to 0 with the switch button pressed, and infinity with the button out.

● Replace the switch if you obtain different test results *(next step)*.

2. UNFASTENING THE LID SWITCH

On most washers, the screws are located in the recess under the lid.

● Lift the washer lid and take out the two screws securing the lid switch *(left)*.

3. REMOVING THE LID SWITCH

● From the underside of the top of the washer housing, detach the switch's green ground wire *(left)*.

● Free the wiring harness from the lip of the housing *(inset)*. On some washers, the harness receptacle is held in place by two spring clips, use a screwdriver to pry off the clips, if necessary.

Agitator and Tub

1. REMOVING THE AGITATOR CAP

If the washer twists clothing excessively, disassemble and inspect the agitator assembly. It is also necessary to remove the agitator to inspect the tub for leaks.

● Lift out the softener dispenser, if any, from the center of the agitator cap.

● With a screwdriver, pry off the cap *(right)*.

● Washers with a softener dispenser in the agitator have a shield with a watertight O-ring covering the agitator bolt *(inset)*. Remove the shield and inspect the O-ring; replace the shield if necessary.

Cap

Softener Dispenser Shield

1

2. REMOVING THE AGITATOR BOLT

● With a socket wrench fitted with an extension, remove the bolt securing the agitator.

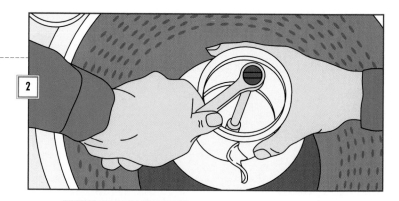

2

3. LIFTING OUT THE AGITATOR

If the agitator dogs, part of the clutch assembly, are worn, the agitator will move in only one direction, twisting clothes and impeding cleaning action.

● To remove the spin drum for tub inspection, first lift out the agitator top, then remove the bottom *(right)*.

● Pull out the clutch assembly from the agitator top and inspect the dogs *(inset)*. Replace them if they are worn, snapping new dogs into place.

Clutch Assembly

Dog

3

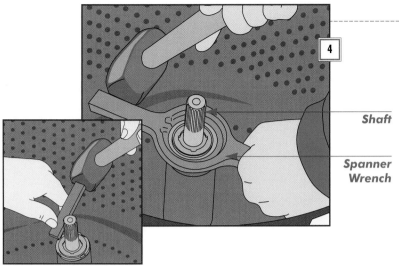
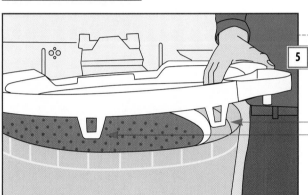

Shaft

Spanner
Wrench

4. DISLODGING THE SPANNER NUT

• Place a spanner wrench over the nut encircling the transmission shaft. Strike the wrench with a hammer or rubber mallet to loosen the nut *(left).* (Take care not to damage the spin basket's porcelain finish.) If you don't have a spanner wrench, substitute a long block of scrap wood *(inset).*

• After freeing the nut, remove it by hand.

5. REMOVING THE SPLASH GUARD

• With your fingers, release the plastic tabs securing the splash guard to the tub. (Don't pry with a screwdriver; doing so may snap the tabs.)

• Lift off the splash guard *(left),* and set it to one side.

Tabs

Ron's TRADE SECRETS

TOUCHING UP A CHIPPED PORCELAIN FINISH

When I spot a chip in the finish of a washer's spin drum, I touch it up right away. Exposed metal rusts quickly—and that leads to rust stains on clothing.

If rust is already present, I sand it off with 400-grit wet-dry abrasive paper—right down to the bare metal. After this, I coat the bare spots with porcelain touch-up paint or an epoxy-based coating—both available at appliance parts dealers and good hardware stores. Some brands of paint come with a small brush, but if not you can simply dab it on with a paper matchstick.

At best, paint touch-ups are temporary. If you plan to keep your washer for several years, the better bet would be to replace the damaged spin drum with a new one.

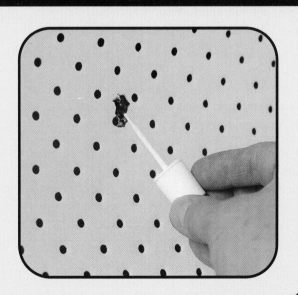

6. REMOVING THE SPIN DRUM

The tub may become worn or cracked from contact with the spin drum caused by load imbalances.

● Gently rock the spin drum back and forth, then turn it a half-rotation. Gently rock the drum again to break any soap deposits that could cause it to stick.

● Lift the spin drum straight up out of the tub *(right)*.

● Inspect the inside of the tub for signs of rubbing or cracks. Call for service if you find any.

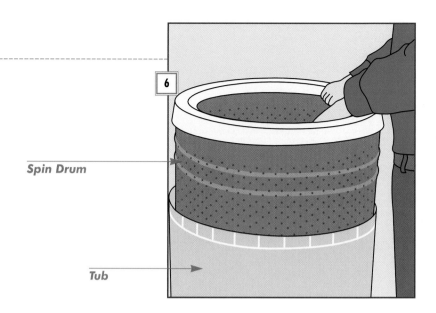

Spin Drum

Tub

Water Inlet Valve

1. DISCONNECTING THE SUPPLY HOSES

● Disconnect power to the washer and turn off the water supply.

● Loosen the couplings with adjustable pliers *(right)*; remove the hoses. Have a bucket on hand to catch dripping water.

● Inside each valve port is a filter screen. Without removing them, check for mineral buildup and clogs. If the filter screens are more than one-third clogged, replace the valve *(Step 3)*.

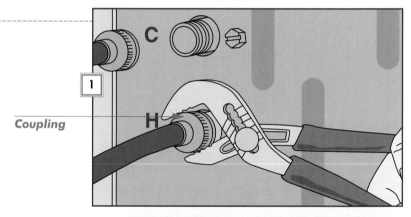

Coupling

2. TESTING THE VALVE SOLENOIDS

● Locate the inlet valve solenoids opposite the hose connectors.

● Label and remove the wires from the valve's solenoid terminals. Set a multitester to RX1 and touch a probe to each pair of terminals in turn *(right)*. The meter should indicate between 100 and 1,000 ohms of resistance on each pair *(page 18)*.

● Replace the valve if the test result is different *(next step)*.

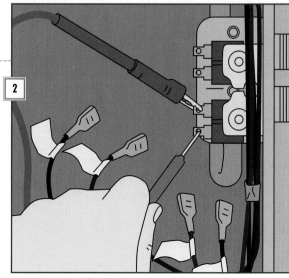

3. REMOVING THE VALVE

• With a nut driver, unscrew the bolts securing the valve to the washer cabinet *(left)*.

• Replace the inlet valve with a new one, then reattach the wires to the solenoid terminals.

Direct-Drive Pump

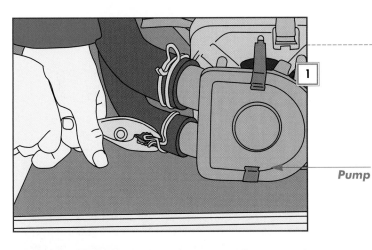

Pump

1. DISCONNECTING THE HOSES

• Unplug the washer, turn off the water supply, and remove the housing *(page 110)*.

• Squeeze the hose clamps with pliers *(left)*, and work them along the hoses, away from the pump. Then wiggle the hoses loose from the pump inlet and outlet.

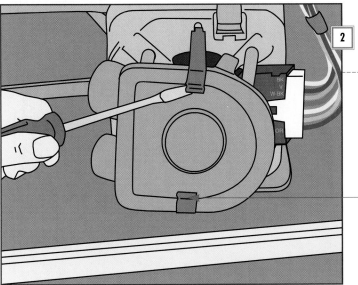

Clip

2. UNSNAPPING THE CLIPS

• With the tip of a screwdriver, pry off the clips securing the pump to the motor *(left)*.

3. REMOVING THE PUMP

• Pull off the pump *(right)*. Examine it for damage and remove any small articles of clothing that may be blocking it.

• Replace the pump if it is damaged or if it continues to leak after you have reinstalled it. (When reattaching the pump, be sure to connect the inlet hose and outlet hose to the correct port.)

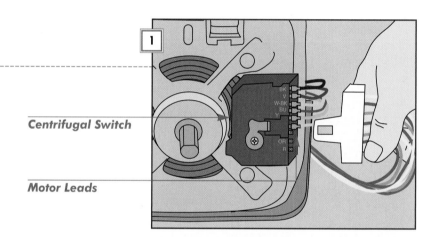

Direct-Drive Motor

1. DISCONNECTING THE CENTRIFUGAL SWITCH

• Disconnect the power to the washer.

• Remove the harness plug attached to the centrifugal switch *(right)*.

• Label and remove the motor leads from the centrifugal switch.

• Remove the mounting screw connecting the centrifugal switch to the motor.

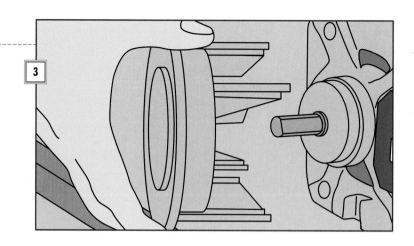

Centrifugal Switch

Motor Leads

2. TESTING THE CENTRIFUGAL SWITCH

• Set a multitester to RX1 and test the switch for the following results: Press the switch lever in *(right)*. The multitester should show continuity between R to BK and OR to BU *(page 18)*. With one probe on R, test all other terminals; none should show continuity.

• Release the switch. The multitester should now show continuity between OR and V. There should be no continuity between the other terminals. Replace the switch if you obtain different test results.

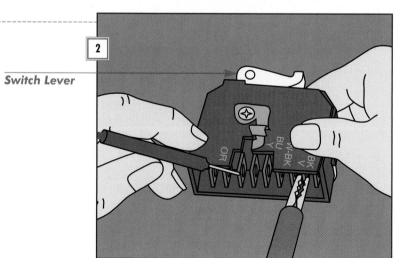

Switch Lever

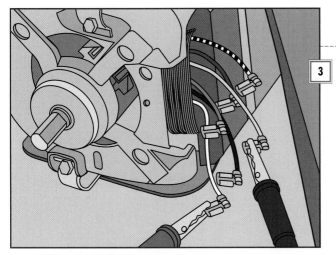

3. TESTING THE MOTOR

• Set a multitester to RX1 *(left)*. The following wire combinations should show continuity: white to blue, white to black/white, white to violet, and yellow to black.

• The following wire combinations should not show continuity: white to yellow, and white to black.

• Replace the motor if you obtain different test results.

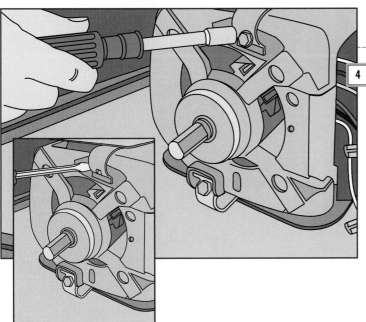

4. REMOVING THE MOTOR

• With a nut driver, unscrew the bolts holding the retaining clips in place *(left)*.

• With the bolts removed, use a screwdriver to pry out the clips securing the motor *(inset)*. Some motors may be secured by bolts only.

• Slide out the motor and mount a replacement.

A COURSE ON COUPLINGS

When the agitator on a direct-drive washer doesn't move as it should, logic dictates having a look at the transmission. When disconnecting the motor from the transmission (to pull it out for inspection), you'll come across the coupling—an assembly of three plastic disks joined by interlocking tabs. A worn coupling causes slippage between the motor and the transmission, sometimes halting the agitator.

Inspect the disks and tabs. If they're worn or broken like the set shown at right, replace the coupling and retest the agitator action. You just might save the trouble and expense of replacing the transmission.

Direct-Drive Transmission and Clutch

1. UNBOLTING THE TRANSMISSION

Suspect transmission problems if the spin drum doesn't spin or if the agitator doesn't agitate.

● Disconnect the power and water supply from the washer. Remove the agitator *(page 114)*, spin drum, and tub *(page 116)*. Disconnect and remove the pump *(page 117)* and motor *(page 119)*.

● With the aid of a helper, lay the washer on its back to expose the transmission from the bottom side.

● With a socket wrench, remove the mounting bolts *(right)*.

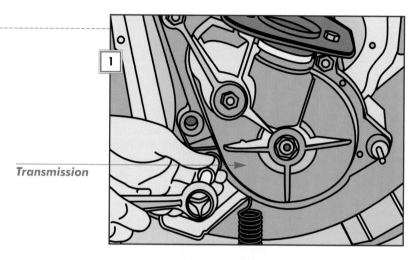

Transmission

2. REMOVING THE TRANSMISSION AND DRIVE SHAFT

● Pull out the transmission *(right)*; the drive shaft will also pull free.

● With a socket wrench, remove the nuts securing the motor mounting plate.

● Take the transmission to a professional for inspection. It is generally considered wiser to replace rather than repair a transmission.

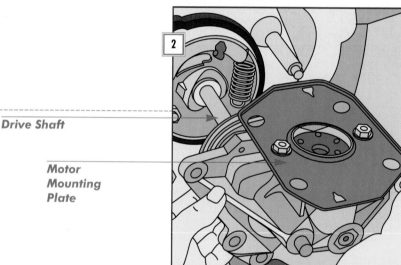

Drive Shaft

Motor Mounting Plate

3. REMOVING THE CLUTCH

If the spin drum won't stop spinning, the clutch may need replacement.

● Pull the thrust bearing off of the drive shaft.

● With a screwdriver, pry off the retaining clip securing the clutch assembly *(right)*. Holding the shaft vertical to prevent spillage of transmission lubricant, slide the clutch off the shaft.

● Reverse the steps for reassembly.

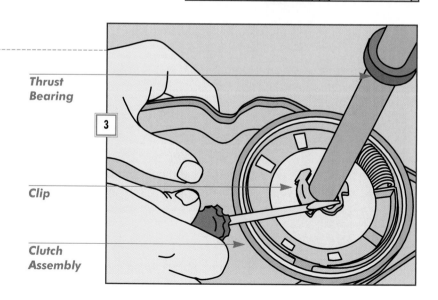

Thrust Bearing

Clip

Clutch Assembly

Drive Belt

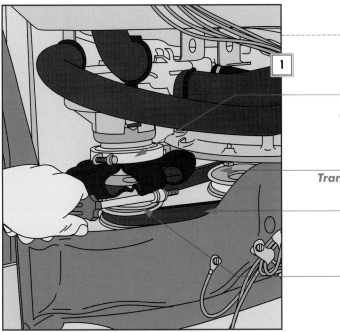

Pump
Coupling
Clamp

Transmission
Pulley

Belt

Clutch
Pulley

1. REMOVING THE DRIVE BELT

• Unplug the washer and remove the back panel *(page 110)*.

• Unscrew the clamps above and below the pump coupling and remove the coupling *(left)*.

• With a socket wrench fitted with an extension, loosen the motor mounting nuts on the mounting plate to release tension on the belt.

• As you pry off the belt, reach under the tub and turn the transmission pulley.

• Remove the belt from the clutch pulley and pull the belt out of the machine.

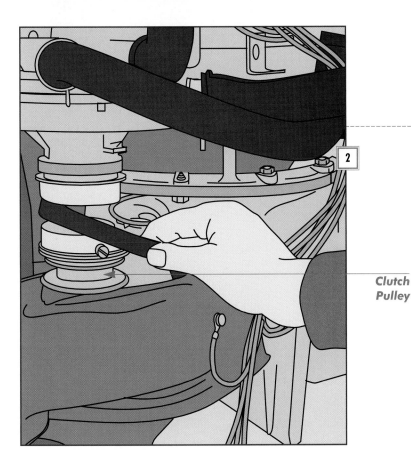

Clutch
Pulley

2. INSTALLING A NEW DRIVE BELT

• Fit the belt around the transmission pulley and hold it there while you loop it around the clutch pulley *(left)*.

• If the belt is tight, rotate the transmission pulley with your hand and shift the motor slightly towards the right.

• Reinstall the coupling and tighten the clamps.

• Push the belt with your thumb; if you can deflect it more than 1/2 inch, pull the motor toward you to take out the slack, then tighten the motor mounting nuts.

FIX IT: Clothes Dryers

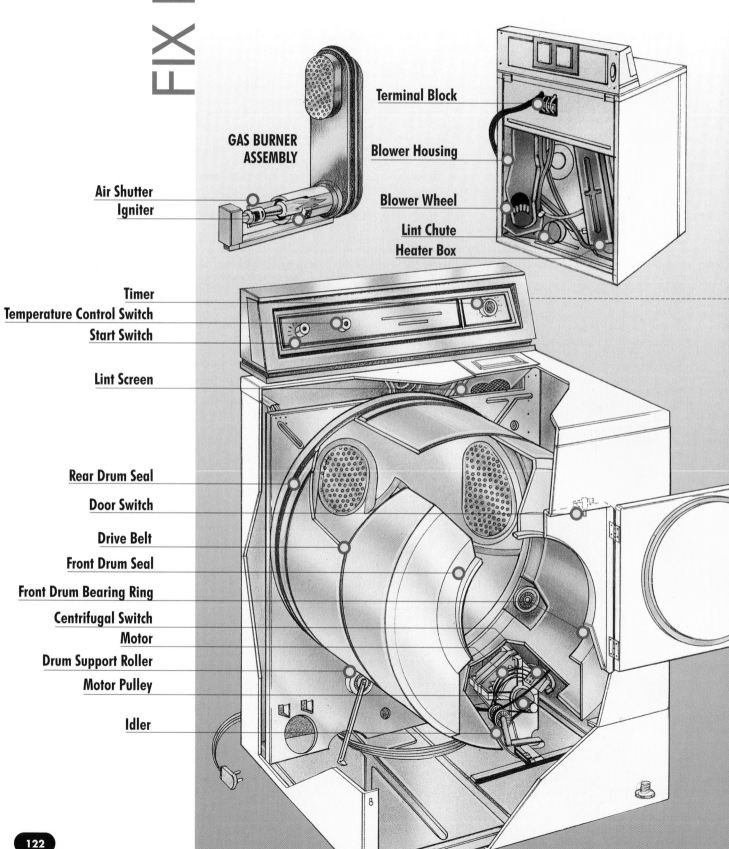

GAS BURNER ASSEMBLY

Air Shutter
Igniter

Terminal Block

Blower Housing

Blower Wheel

Lint Chute

Heater Box

Timer
Temperature Control Switch
Start Switch

Lint Screen

Rear Drum Seal

Door Switch

Drive Belt

Front Drum Seal

Front Drum Bearing Ring

Centrifugal Switch
Motor

Drum Support Roller

Motor Pulley

Idler

Chapter 7

Contents

How They Work

The clothes dryer shown here is an electric model. It differs from the gas-fired variety mainly in the source of heat. All dryers combine air, heat, and motion to evaporate moisture from clothes. A motor turns a drive belt that revolves a drum, which tumbles wet garments to prevent them from clumping together. At the same time, a blower, also powered by the motor, forces air past a heating element or a gas flame *(upper left)* and into the drum. The flow of warm air draws lint and moisture from the clothes through a lint screen and out the exhaust duct. Various switches—including a centrifugal switch that interrupts heat to the dryer should the drum happen to stop—thermostats, and a timer regulate temperature, drying time, and other cycle variations.

Troubleshooting

Problem	Solution
• **Dryer doesn't run at all**	Check that the dryer is plugged in; check for a blown fuse or circuit breaker • Inspect the power cord • Test the door switch **132** • Test the start switch **129** • Test the timer and timer motor **131** • Test the motor **140** •
• **Motor runs, but dryer doesn't heat**	Check for a blown fuse or circuit breaker • Test the temperature selector switch **130** • Test the timer and timer motor **131** • Test the thermostats **133** • Test the centrifugal switch **131** • Replace the heating element **138** •
• **Motor runs, but drum doesn't turn**	Check the drive belt **134** • Check the idler **134** • Service the drum **136** •
• **Dryer runs with door open**	Test the door switch **132** •
• **Dryer doesn't turn off**	Room too cool; room temperature must be at least 50°F for dryer to work • Test the timer and timer motor **131** • Test the heating element **138** • Test the thermostats **133** •
• **Drying time is too long**	Clean lint screen, exhaust duct, and vent • Test the thermostats **133** • Replace the heating element **138** •
• **Drying temperature is too hot**	Clean lint screen; clean or unkink exhaust duct and vent • Test the thermostats **133** • Replace the heating element **138** •
• **Dryer is noisy**	Dryer not level; adjust leveling feet • Tighten loose screws on panels and trim • Check the drive belt **134** • Check the idler **134** • Check the drum seals **136** • Check the drum support rollers **136** • Service the blower **140** •

Before You Start

Many common repairs are similar for all types of dryers. Where differences exist among the many models, they mostly concern the source of heat.

GAS VERSUS ELECTRICITY

In electric dryers, diagnosing heating-element problems and solving them is something almost anyone can do. However, when heat fails in a gas dryer, it's best to call for service. Aside from adjusting the air shutter to optimize the flame, there's little to repair that doesn't require special training. There's usually not even a pilot flame to light— most gas dryers have an electric ignition.

Once past the source of heat, all dryers are simple machines and remarkably alike. Procedures for repairing one model generally apply to all others as well.

A PERVASIVE GREMLIN

Lint is a dryer's worst enemy. Even if you clean the filter after every load, lint accumulates around the moving parts, as well as inside the exhaust duct and vent, forcing the machine to work harder. At least once a year, turn off the power to the dryer, remove the front and rear panels, and vacuum or brush lint from the motor, idler, and gas burner, if any.

Before You Start Tips:

···: If your electric dryer runs but doesn't heat, check the two dryer-circuit fuses or circuit breakers at the service panel. If one has blown or tripped, it can interrupt heat without affecting the motor.

TOOLS

Screwdriver
Putty knife
Nut driver
Multitester
Long-nose pliers
Adjustable wrenches

MATERIALS

Masking tape
Scrap wood
Rags
Paint thinner
Rubber adhesive

SAFETY FIRST

If you have a gas dryer, don't risk rupturing the gas line by moving the dryer or disconnecting the gas yourself. Call the gas company or a service technician to disconnect and move the dryer for you.

Access to the Controls

1. FREEING THE CONTROL CONSOLE

• Unplug the dryer or turn off power to the dryer at the main service panel.

• Unscrew the console at each end. Some models have screws at the bottom front of the console *(right)*. Others have screws at the top or sides of the console. To expose the screws you may have to peel back a decorative strip that conceals them.

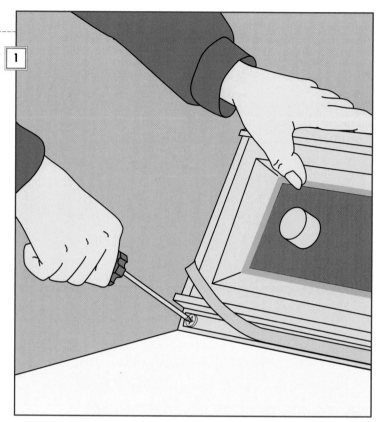

2. EXPOSING THE CONNECTIONS

• Spread a towel on top of the dryer to protect its finish.

• Roll the console facedown onto the towel *(right)*. On some models you must first slide the console forward to disengage tabs on the end panels from slots in the dryer top.

• If the console has a rear panel, unscrew it to expose the start switch, temperature selector, circuit diagram, and timer.

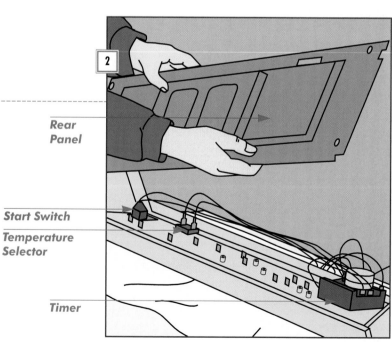

Rear Panel

Start Switch

Temperature Selector

Timer

The Top

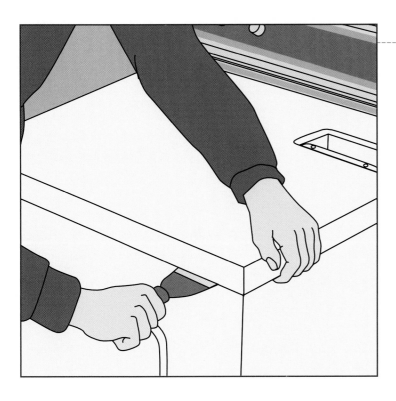

UNLATCHING THE CATCHES

Raise the top to gain access to the drum.

• On an electric dryer, which has a top-mounted lint screen, unplug the dryer, pull out the screen, and remove the two screws at the front edge of the screen slot.

• Insert a putty knife wrapped in masking tape under the top, then push in to disengage the hidden clips securing the top in place *(left)*.

The Rear Panel

UNFASTENING THE REAR PANEL

• Unplug the machine, disconnect the exhaust duct, and move the dryer away from the wall.

• Remove the screws around the panel edges *(left)*, then take the panel off. Some models have two or three smaller panels; remove each as needed.

CAUTION: If you have a gas dryer that must be disconnected from the gas supply in order to move it, call the gas company or a service technician to disconnect the dryer for you.

The Toe and Front Panels

1. REMOVING THE TOE PANEL

• Unplug the dryer and remove any retaining screws.

• Insert a putty knife near the center top of the toe panel.

• Push down and in against the hidden clip while pulling on a corner of the panel *(right)*.

• Lift the panel off the two bottom brackets and remove it.

2. UNSCREWING THE FRONT PANEL

• Raise the top *(page 127)* and remove the toe panel *(above)*.

• Loosen, but do not remove, the screws (if any) at the bottom front corners of the front panel *(right)*.

• Support the drum with blocks of scrap wood so that it won't drop down when you remove the front panel.

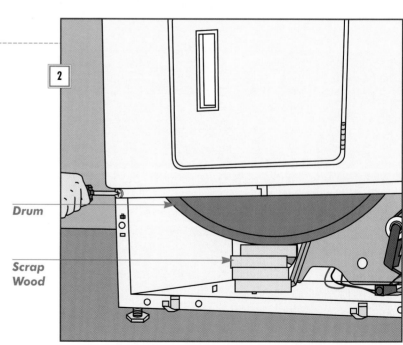

Drum

Scrap
Wood

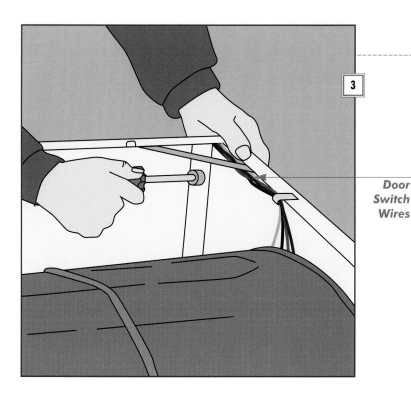

3. REMOVING THE CORNER SCREWS

• Move to the inside of the dryer housing. Disconnect and label the wires leading to the door switch.

• Supporting the front panel with one hand, remove the screws at each inside corner *(left)*.

• Lift the panel off the lower screws or brackets.

Door Switch Wires

Start Switch

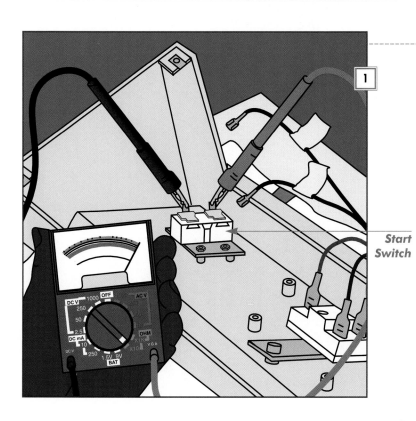

1. CHECKING FOR CONTINUITY

• Unplug the dryer. Free the control console and tilt it forward *(page 126)*.

• Label and disconnect the wires from the start switch terminals.

• To test a two-terminal switch *(left)*, set a multitester to RX1. Clip one probe to terminal CO (or R2) and the other to NO (or R1). The meter should indicate an open circuit before you press START, and continuity when you press it *(page 18)*.

• For a three-terminal switch, set a multitester to RX1 and clip one probe to terminal NC (or CT1) and the other to CO (or R1); the meter should indicate an open circuit when you press the START button, and continuity when you release it.

• Replace the switch if it fails either test *(next step)*.

Start Switch

2. REMOVING THE START SWITCH

• Pull off the control knob and unscrew the switch from the console *(right)*.

• Remove and reuse the switch mounting bracket, if any.

• Screw the new switch in place and reconnect the wires.

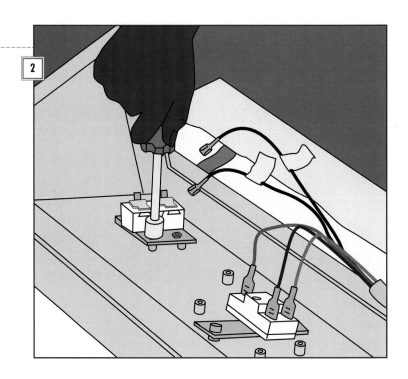

Temperature Selector Switch

TESTING THE SWITCH

Rotary and push-button switches *(inset)* are tested the same way. Unless you can read the wiring diagram, located on the back of the dryer or inside the control console, tests on switches are best left to a professional.

• Unplug the dryer. Free the control console and tilt it forward *(page 126)*. Label and disconnect the wires from the switch.

• Check the wiring diagram to identify terminals regulating the inoperative cycle.

• Set the knob to the inoperative cycle. Set a multitester to RX1 and touch one probe to each terminal *(right)*. Look for continuity.

• To replace a faulty switch, unscrew the old switch from the control console and install a new switch, transferring the wires.

Temperature Selector Switch

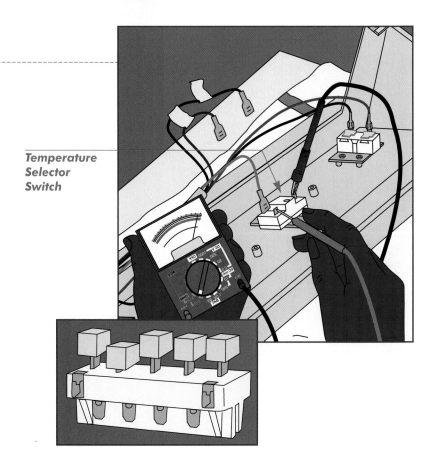

Timer

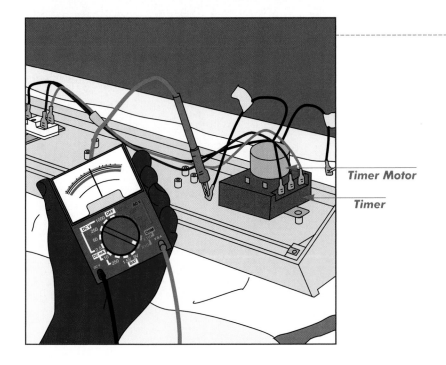

Timer Motor

Timer

CHECKING THE MOTOR

Checking the timer motor is a simple job that anyone can do. However, unless you can read the wiring diagram, located on the back of the dryer or pasted inside the control console, checking other aspects of timer operation is best left to a professional.

• Unplug the dryer. Free the control console and tilt it forward *(page 126)*.

• Disconnect the two wires leading to the timer motor. Set a multitester to RX1000. Touch one probe to each lead on the motor wires *(left)*; the meter should read 2,000 to 3,000 ohms of resistance *(page 18)*.

• If you obtain a different result, replace the timer motor.

Centrifugal Switch

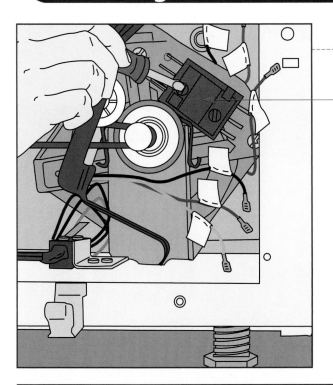

Centrifugal Switch

MULTIPLE CONTINUITY TESTS

The centrifugal switch is mounted on the motor. The method of gaining access to it varies among models.

• Disconnect and label the wires, then unscrew the switch from the motor *(left)*.

• Set a multitester to RX1000 and place the probes on terminals 1 and 2, then 5 and 6, then 5 and BK (or 3).

• Test with the switch button out, then in, checking for continuity and resistance *(page 18)*. Replace the switch if test results differ from those listed in the table below the illustration at left.

TERMINAL PAIRS	1–2	5–6	5-BK(3)
Out	Continuity	Continuity	Resistance
In	Resistance	Resistance	Continuity

Door Switch

CHECKING FOR CONTINUITY

• Raise the dryer top *(page 127)* to reach the door switch, which is mounted near one of the upper corners of the dryer front.

• Disconnect the wires from the terminals and set a multitester to RX1. Touch a probe to each terminal that was connected to a wire *(right);* ignore any other terminals.

• When the door is closed, the meter should show continuity; when the door is open, the meter should indicate an open circuit *(page 18).*

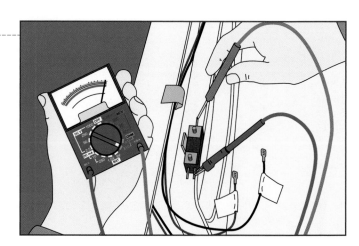

REMOVING A LEVER SWITCH

• Remove the screws on either side of the lever *(right).*

• Lift out the lever switch from the side of the dryer.

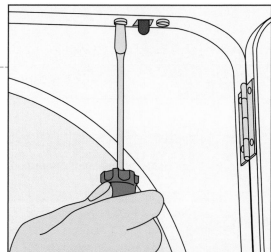

Ron's TRADE SECRETS

GAS SUPPLY-HOSE ALERT

If you haven't checked the hose that connects your gas dryer to the supply line in your home for some time, do it now. Chances are you'll find an old-style flexible steel hose, which has a dull or brushed finish, or a hose that has a gray plastic coating.

If you find either type, my advice is to get it replaced with a modern stainless steel hose *(right).* Stainless steel is immune to the water vapor that condenses from gas and rusts ordinary steel supply hoses from within. The resulting leaks can be fatal.

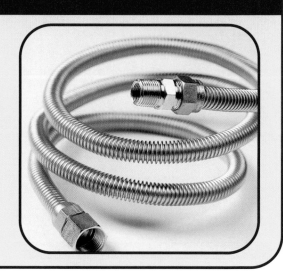

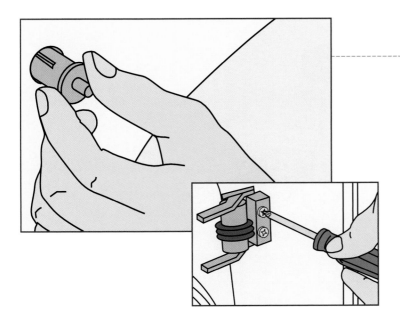

DETACHING OTHER DOOR SWITCHES

• To remove a push-button switch *(left)*, reach down inside the dryer. Squeeze the retainer clips on the back of the switch, then pull the switch out through the front.

• To gain access to a hinge-mounted switch *(inset)*, take off the front panel and unscrew the switch from the drawer hinge.

Thermostats and Fuses

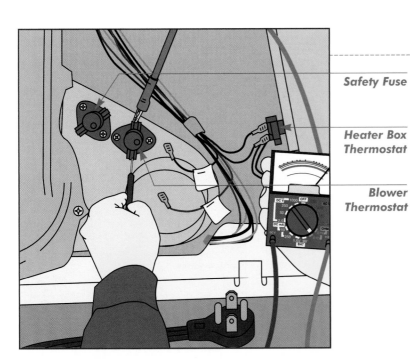

Safety Fuse

Heater Box
Thermostat

Blower
Thermostat

TESTING THE SWITCH INSIDE

The safety fuse and thermostats are located on the blower housing and on the heater box. Test each the same way.

• Unplug the dryer. Disconnect the wires from the thermostats and safety fuse, one at a time, and label their positions.

• Set a multitester to RX1 and touch a probe to each terminal *(left)*; the meter should show continuity *(page 18)*.

• Replace any thermostat or fuse that fails the test.

Drive Belt and Idler Pulley

1. DISENGAGING THE BELT

To reach the idler pulley, remove the top panel, or raise the top and remove the front panel *(pages 126-128).*

• Prop the dryer drum on a piece of scrap wood.

• Push the idler pulley toward the motor pulley to slacken the drive belt, then disengage the belt *(right).*

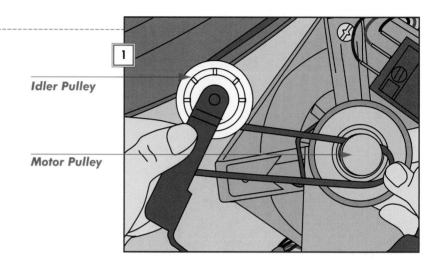

Idler Pulley

Motor Pulley

2. REMOVING THE IDLER

• Inspect the idler bracket, pulley, and spring. Idlers vary in style; many are one piece and are held in place in the dryer floor by belt tension *(top right).*

• Another type of idler has a tension spring *(bottom right).* Unhook the spring and replace it if it is worn or broken. This type of idler may have a replaceable pulley *(Step 3).*

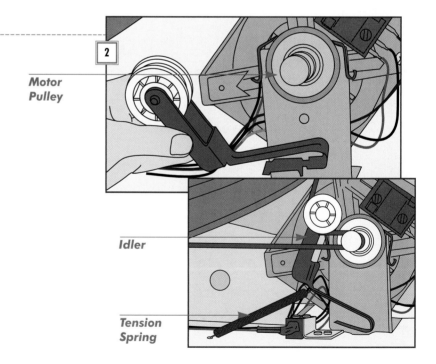

Motor Pulley

Idler

Tension Spring

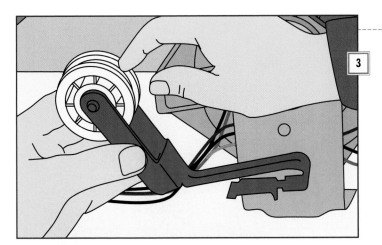

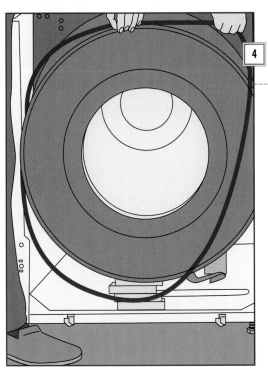

3. CHECKING THE IDLER PULLEY

• Inspect the surface of the pulley for uneven wear and move it back and forth to check for wobbling *(left)*.

• Use a nut driver to remove the screw (or long-nose pliers to remove a retaining ring) at one end of the axle, then slide the axle out of the pulley.

• Place a new pulley and washers in the bracket and insert the axle. Replace the screw or retaining ring.

4. REMOVING THE DRIVE BELT

• Lift the drum slightly and slide the loose belt free *(left)*.

• Align a new belt in the same position as the old one, with its grooved side against the drum.

• To rethread the belt, push a loop of the belt under the idler pulley and catch it on the motor pulley. Check that the rear drum seal rides properly against the bulkhead behind the drum *(page 136)*.

RECOGNIZING A FAULTY DRIVE BELT

It's a good idea to occasionally inspect dryer drive belts for wear. Your dryer may have an access panel for this purpose. If not, have a look at the belt whenever you remove the front panel to clean lint build-up.

Examine the belt for cracks and brittleness. A belt in good condition has a smooth, rubbery feel. Healthy dryer belts also have grooves running around the inner surface. When the grooves are missing in sections, as in the photograph at the right, the belt is ready to fail.

The Drum

1. CHECKING AND REPLACING THE FRONT DRUM SEAL

• Unplug the dryer, lift the top *(page 127)*, and remove the toe panel, if any, and the front panel *(page 128)*.

• Inspect the felt seal surrounding the door opening on the back of the front panel. Look for objects embedded in the felt.

• Peel off the old seal. Position a new seal on the flange *(right)*, securing it with a bead of the heat-resistant rubber adhesive sold with the seal.

• Check the plastic bearing ring in the drum opening. If it is rough or worn, remove it by releasing the plastic retaining clips. Then fit a new ring to the drum.

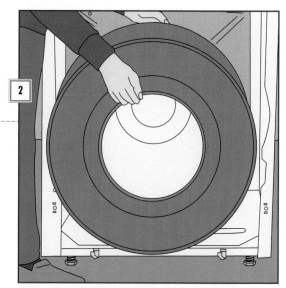

Bearing Ring

Felt Seal

2. REMOVING THE DRUM

• Disengage and remove the drive belt *(page 134)*.

• Lift the drum slightly—it weighs about 5 pounds—then slide it through the front of the cabinet *(right)*.

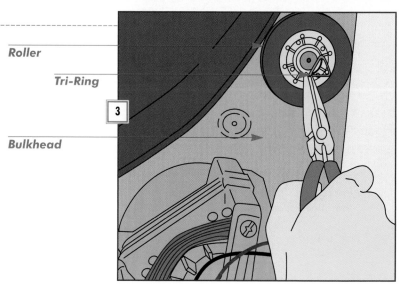

3. REPLACING DRUM SUPPORT ROLLERS

• Locate the two rubber rollers mounted on the bulkhead supports. Check each roller for wear and replace it if it is damaged or wobbles back and forth.

• To do so, use long-nose pliers to pry off the tri-ring that secures the roller to the shaft *(right)*, then slide off the roller. With pliers, pull off the support bracket from the left roller (not shown).

• Slide on a new roller, and pop the tri-ring back in place. Press the bracket on the left roller back onto the shaft.

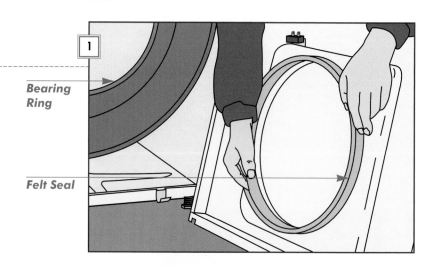

Roller

Tri-Ring

Bulkhead

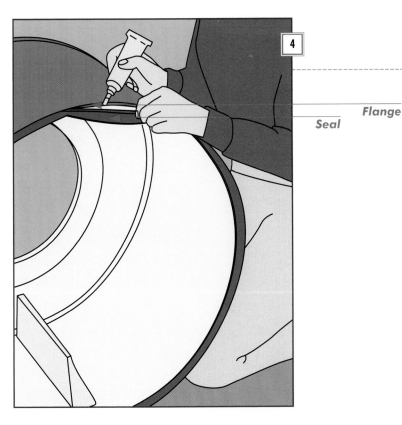

Flange

Seal

4. REPLACING A REAR DRUM SEAL

• Check the felt-and-plastic seal around the drum's back edge; if the seal is damaged, scrape it off with a putty knife.

• Clean any adhesive off the drum flange with a rag and paint thinner (do not use lacquer thinner).

• Slip a new seal around the drum flange with its stitched edge in.

• Lift the inner edge of the seal and apply around the drum flange a bead of the heat-resistant rubber adhesive, pressing the seal down as you go *(left)*.

• Let the adhesive set for one hour.

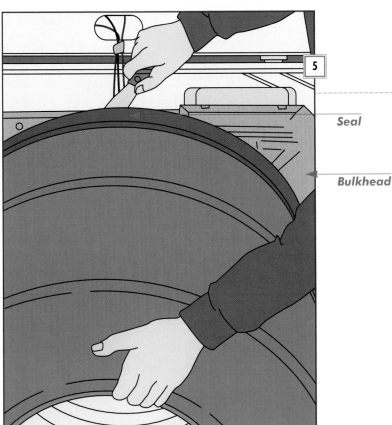

Seal

Bulkhead

5. REINSTALLING THE DRUM

• Slide the drum in through the front of the dryer and temporarily rest the drum flange on the support rollers.

• Rethread the drive belt *(page 134)*.

• Seat the rear seal against the bulkhead by inserting a putty knife between the seal and bulkhead *(left)*. Rotate the drum a full rotation to be sure that the seal edge is not pinched by the rollers.

• Replace the dryer panels.

Electric Dryer Heating Elements

1. CHECKING RESISTANCE

● Unplug the dryer and remove the rear panel *(page 127)*.

● Label and disconnect the wires to the heater terminals.

● Set a multitester to RX1. If the heater has two terminals, touch one probe to each terminal *(right)*. The meter should show 5 to 50 ohms *(page 18)*.

● If the heater has three terminals, touch one probe to the middle terminal and the other probe to the outer terminals in turn. The meter should read 10 to 40 ohms in each case *(page 18)*.

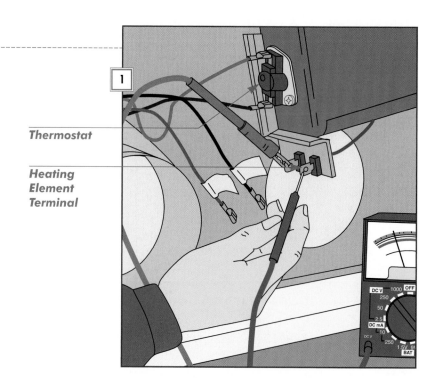

Thermostat

Heating Element Terminal

2. TESTING FOR GROUND

● Set a multitester to RX1 and touch one probe to the heater box *(right)*.

● Touch the other probe to each terminal in turn (whether there are two or three). The meter should not move *(page 18)*.

● If the heating element fails this test or the one in Step 1, replace it *(next page)*.

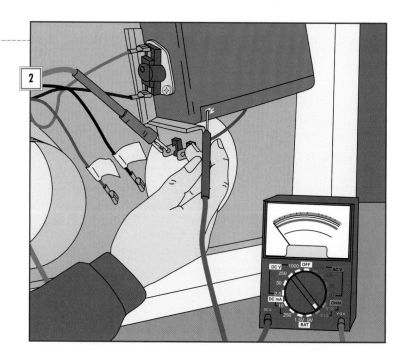

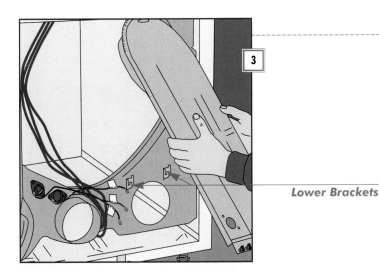

Lower Brackets

3. REMOVING THE HEATER BOX

- Raise the dryer top *(page 127)*.

- Label and disconnect the thermostat wires. Remove the thermostat from the heater box, followed by the screws holding the heater box to its lower support brackets.

- Lift the heater box slightly to free it from the lower brackets, then pull it down and out from the rear of the dryer *(left)*.

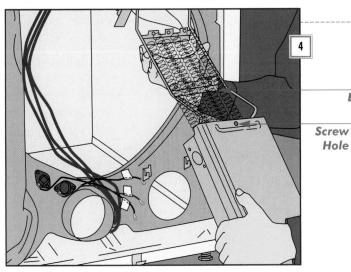

Element

Screw
Hole

4. INSTALLING A NEW ELEMENT

- Remove the screw holding the element in the heater box, then carefully pull out the element, noting its orientation *(left)*.

- Slide a new element into the box, same side up as the original. Be sure that the coils do not rub against the side of the box. Insert and tighten the screw.

- Slide the heater box into the rear of the dryer; hook the slots into their brackets.

- Screw the thermostat back on the heater box, reconnect all of the wires, and replace the rear panel.

SERVICING OTHER DRYER HEAT SOURCES

Some electric dryers have circular heating coils *(near right)*. To replace them, flatten the small tabs with pliers, then pull the terminals through the block and remove the old coil. Stretch the new coil to the same length before installing it.

Gas dryer burners seldom fail. More often, problems occur in the electronic ignition or pilot assembly. Leave these repairs to a professional. But you can adjust the air shutter to admit more or less air: Loosen the thumbscrew and rotate the shutter until the flame is light blue *(far right)*.

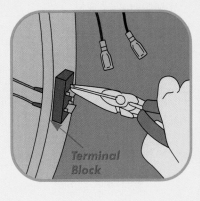

Terminal
Block

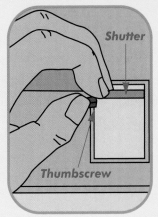

Shutter

Thumbscrew

Motors and Blowers

1. Testing the motor

Unless you can read the wiring diagram, located on the back of the dryer or pasted inside the control console, proper testing of the motor is best left to a professional technician. However, this step details a common configuration.

• Unplug the dryer and gain access to the motor *(pages 126-128)*.

• Label and disconnect the wires attached to the centrifugal switch. Set a multitester to RX1. Connect the probes to the yellow and blue wires leading to the motor *(right)*. The meter should read 1 to 5 ohms of resistance *(page 18)*.

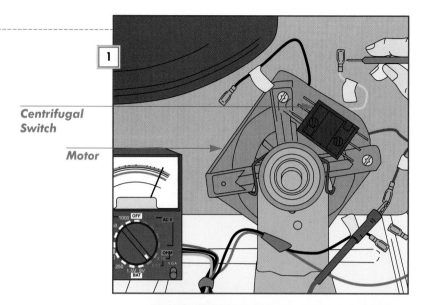

Centrifugal Switch

Motor

2. Releasing the blower wheel

• With two wrenches, grip the motor shaft in front of the motor and the blower wheel hub behind the motor *(right)*. If necessary for access, remove the drum *(page 136)*.

• Hold the blower wheel stationary as you turn the motor shaft clockwise toward the side of the dryer. Turn until the wheel is free of the shaft.

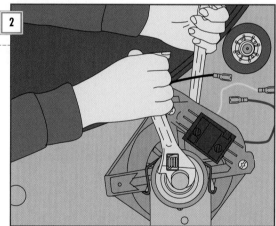

3. Inspecting and replacing the blower

• Remove the rear panel *(page 127)*. Unscrew the lint chute and slide it aside to expose the blower housing *(right)*.

• Unscrew the blower wheel from the motor shaft and examine it for damage to the fins or threads. If you find any, replace it.

• Install the blower wheel now or, if you plan to remove the motor, wait until Step 5. Begin by loosely threading the hub of the blower onto the motor shaft.

• Tighten the shaft from inside the dryer, using the same double-wrench technique described in Step 2.

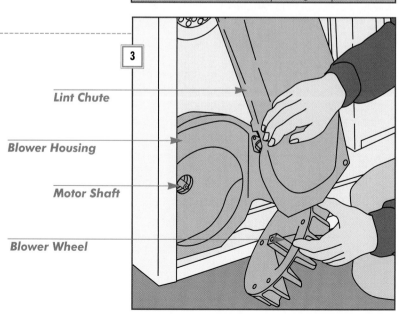

Lint Chute

Blower Housing

Motor Shaft

Blower Wheel

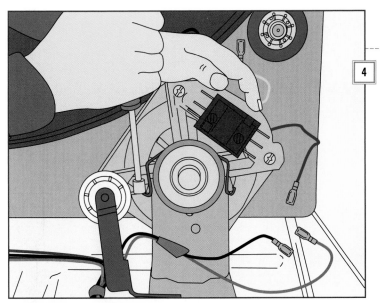

4

4. RELEASING THE MOTOR CLAMPS

• Remove the belt *(page 134).*

• Label the green ground wire (not shown), then unscrew it from the motor housing.

• Locate the straplike spring clamps that secure the round rubber cushions at the front and back of the motor, hidden in this view, to the motor bracket. Press down on the clamp's hooked end with a nut driver and snap it off the motor bracket *(left).*

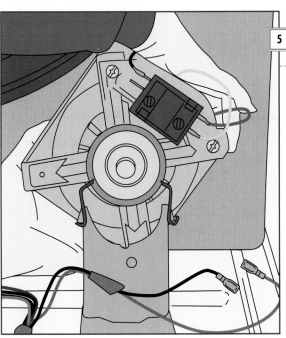

5

5. REINSTALLING THE MOTOR

• Set the rubber cushions in the motor brackets, fitting the tab on the front cushion into the slot in the front bracket.

• Position the motor with the threaded end of the shaft to the rear of the dryer *(left).*

• Place the clamps across the rubber cushions and snap them onto the brackets with a nut driver.

• Thread the blower onto the rear motor shaft *(Step 3).*

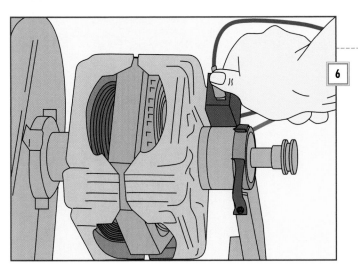

6

6. RECONNECTING WIRES TO THE MOTOR

• Reconnect the wires to the centrifugal switch *(left),* taking care to screw the green ground wire to the motor housing.

• Install the drum, idler, and belt, and replace the dryer panels.

Index

Ovens,
Electric, 41, 43, 51
Gas, 61, 70, 71
Refrigerators, 24
See also Temperature
controls; Thermostats

I

Icemakers:
Modular, 32
Access, 32
Motor, 32
Heater, 32
Shutoff arm, 32
Overflow, 35
Thermostat, 32, 33
Water inlet valve, 32, 39
Non-modular, 34
Access, 34, 35
Ejector gear, 35
Motors, 34, 37
Overflow, 35
Shutoff arm, 34, 35, 36
Switches, 34
Holding, 34,36
ON/OFF, 34, 35, 36
Valve, 34, 37
Thermostat, 34, 37, 38
Water inlet valve, 34, 37, 39
See also Freezers;
Refrigerators

L

Lights:
Freezers, 29
Ovens, 41, 55
Refrigerators, 10, 11, 17

M

Motors:
Clothes dryers, 123,
140, 141
Clothes washers, 103,
106,118
Dishwashers, 75, 92
Garbage disposers, 95, 98
Icemakers,
Modular, 32
Non-modular, 34, 37
Multitesters, 18

O

Ovens:
Electric, 41
Burner elements, 41,
43, 51
Cleaning, 43, 52

Door, 41, 56
Gaskets, 41, 43, 57, 58
Light, 41, 55
Selector switches, 55
Self-cleaning, 43, 51, 52
Temperature controls,
52, 54
Thermostat, 41, 53
Wall, 46
Gas, 63 *See also* Ovens,
electric
Burners, 61, 70, 71
Cleaning, 61, 70, 72
Electric igniters, 72
Flame adjustment, 71
Glowbars, 61, 73
Pilots, 69
Thermostat, 61, 70, 73
See also Electric ranges;
Gas ranges

P

Pilots:
Clothes dryers, 139
Gas ranges, 61, 63, 65
Ovens, 69
Power supplies:
Clothes dryers, 125
Dishwashers, 78, 81
Electric ranges, 41, 44
Garbage disposers, 100
Gas ranges, 64
Refrigerators, 11
Pumps:
Clothes washers, 103, 106,
117
Dishwashers, 75, 92

R

Ranges. *See* Electric ranges; Gas
ranges; Ovens
Refrigerators, 6
Access, 22, 27, 28
Breaker strips, 21
Cleaning, 10, 11, 12, 16, 25
Compressor, 7, 28
Condenser, 7, 10, 11, 27
Defrost system, 10
Heater, 24
Timer, 25
Doors, 10, 12, 14, 15
Drains, 10, 11, 12, 25
Evaporator, 7, 10, 11, 23
Frost build-up, 11, 13
Frost-free, 7
Gaskets, 10, 12, 16
Light, 10, 11, 17

Power failures, 11
Power supply, 11
Temperature controls, 10, 19
Troubleshooting, 8
See also Freezers; Icemakers

S

Stoves. *See* Electric ranges;
Gas ranges; Ovens
Switches. *See* Wall switches

T

Temperature controls:
Clothes dryers, 130
Clothes washers, 112
Electric ranges, 41, 48, 50
Freezers, 10, 19
Ovens,
Electric, 52, 54
Refrigerators, 10, 19
See also Heating elements
Thermostats:
Clothes dryers, 123, 132
Icemakers,
Modular, 32, 33
Non-modular, 34,
37, 38
Ovens,
Electric, 41, 53
Gas, 61, 70, 73
See also Heating elements;
Temperature controls
Timers:
Clothes dryers, 123, 131
Clothes washers, 103,
111, 112
Dishwashers, 82
Refrigerators, 25

W

Wall switches, 101
Washers. See Clothes
washers; Dishwashers
Water inlet valves:
Clothes washers, 103, 116
Dishwashers, 75, 90
Icemakers,
Modular, 32, 39
Non-modular, 34, 37, 39

TIME® LIFE

Time-Life Books
is a division of Time Life Inc.

Time Life Inc.

George Artandi
President and CEO

Time-Life Books

Stephen R. Frary
President

Neil Kagan
Publisher/Managing Editor

Steven A. Schwartz
Vice President, Marketing

How To Fix It:

Major Appliances

Lee Hassig
Editor

Kate McConnell
Art Director / Series Designer

Wells P. Spence
Director of Marketing

Monika D. Lynde
Page Make-Up Specialist

Patricia Bray
Special Contributor (design)

Christopher Hearing
Director of Finance

**Marjann Caldwell
Patricia Pascale**
Directors of Book Production

Betsi McGrath
Director of Publishing Technology

John Conrad Weiser
Director of Photography
and Research

Barbara Levitt
Director of Editorial Administration

Marlene Zack
Production Manager

James King
Quality Assurance Manager

Louise D. Forstall
Chief Librarian

Butterick Media

Staff for Major Appliances

Michael Chotiner
Editor

Mark Feirer
Project Development

Caroline Politi
Director of Book Production

Daniel Newberry
Associate Editor

Linda Greer
Associate Editor

Ben Ostasiewski
Art Director

David Joinnides
Page Layout

Jim Kingsepp
Technical Consultant

Annemarie McNamara
Copy Editor

Naomi Bibbins Bain
Editorial Coordinator

Lillian Esposito
Production Editor

Art Joinnides
President

Picture Credits

Fil Hunter
Cover Photograph

**Bob Crimi
Geoff McCormack
Linda Richards
Joseph Taylor**
Illustration

**Brian Kraus
Juan Rios
"Butterick Media"**
Interior Photographs

First printing. Printed in U.S.A.
Published simultaneously in Canada.
School and library distribution
by Time-Life Education,
P.O. Box 85026, Richmond, Virginia 23285-5026.

TIME-LIFE is a trademark of Time Warner Inc. U.S.A.

**Library of Congress
Cataloging-in-Publication Data**
Major Appliances/ by the editors of Time-Life Books.
    p. cm. -- (How to Fix It)
    Includes index.
    ISBN 0-7835-5651-9
1. Household appliances---Maintenance and repair
---Amateurs' manuals.
 I. Time-Life Books. II. Series.
TX298.M28 1998   98-4231
683'.88--dc21   CIP